THE MOLE CONCEPT IN CHEMISTRY

Selected Topics in Modern Chemistry

SERIES EDITORS

Professor Harry H. Sisler
University of Florida
Gainesville, Florida

Professor Calvin A. VanderWer
Hope College
Holland, Michigan

BREY–*Physical Methods for Determining Molecular Geometry*
CHELDELIN AND NEWBURGH–*The Chemistry of Some Life Processes*
EYRING AND EYRING–*Modern Chemical Kinetics*
HILDEBRAND–*An Introduction to Molecular Kinetic Theory*
KIEFFER–*The Mole Concept in Chemistry*
MOELLER–*The Chemistry of the Lanthanides*
MORRIS–*Principles of Chemical Equilibrium*
MURMANN–INORGANIC COMPLEX COMPOUNDS
O'DRISCOLL–*The Nature and Chemistry of High Polymers*
OVERMANN–*Basic Concepts of Nuclear Chemistry*
ROCHOW–*Organometallic Chemistry*
RYSCHKEWITSCH–*Chemical Bonding and the Geometry of Molecules*
SISLER–*Electronic Structure, Properties, and the Periodic Law*
SISLER–*Chemistry in Non-Aqueous Solvents*
SONNESSA–*Introduction to Molecular Spectroscopy*
STRONG AND STRATTON–*Chemical Energy*
VANDERWERF–*Acids, Bases, and the Chemistry of the Covalent Bond*
VOLD AND VOLD–*Colloid Chemistry*

THE MOLE CONCEPT IN CHEMISTRY

WILLIAM F. KIEFFER

Professor of Chemistry
College of Wooster
Wooster, Ohio

VAN NOSTRAND REINHOLD COMPANY
NEW YORK   CINCINNATI   TORONTO   LONDON   MELBOURNE

QD
461
K4

Van Nostrand Reinhold Company Regional Offices:
New York Cincinnati Chicago Millbrae Dallas

Van Nostrand Reinhold Company International Offices:
London Toronto Melbourne

Copyright © 1962 by LITTON EDUCATIONAL PUBLISHING, Inc.

Library of Congress Catalog Number 62-15248

All rights reserved. No part of this work covered by the
copyright hereon may be reproduced or used in any form
or by any means—graphic, electronic, or mechanical,
including photocopying, recording, taping, or information
storage and retrieval systems—without written permission of
the publisher. Manufactured in the United States of America.

Published by Van Nostrand Reinhold Company
450 West 33rd Street, New York, N. Y. 10001

Published simultaneously in Canada by
Van Nostrand Reinhold Ltd.

10 9 8 7 6 5

Series Editors Statement

There are few topics which the chemistry student finds more difficult to understand than the concept of the "mole," yet its mastery is absolutely essential to sound chemical reasoning. Moreover, few chemistry textbook authors permit themselves the space to examine this concept and its manifold implications in a sufficiently thorough fashion to satisfy the inquiring student.

It is, therefore, with great satisfaction and pride that we present "The Mole Concept in Chemistry" as an impressive addition to SELECTED TOPICS IN MODERN CHEMISTRY. The author Professor William F. Kieffer is an experienced, inspiring, and thoroughly sound teacher of college chemistry and, in his capacity as editor of the *Journal of Chemical Education*, is at the forefront of that group of progressive chemical educators responsible for the rapid development of chemical education during the past two decades.

Professor Kieffer has done an outstanding job in treating the mole concept in a thorough and exciting manner. We believe that chemistry teachers will find that this new text fills a vital need for their students.

Harry H. Sisler
Calvin A. VanderWerf

AN OPEN LETTER

Dear Chemistry Student:

The preface of a book often is the last part you, the user, will read. Naturally, an author wishes this were not so. I do too. Consequently, I am writing you, a chemistry student, this letter. Please let me talk to you a few minutes.

This book was written for you, not for your teacher, quiz-section instructor, or professor. There is nothing in these pages that he doesn't know already. In fact there probably is little information that you are meeting for the first time. Why then did I write it?

In one sense I have done a job that you should have done. At least it is typical of the kind of job you should do for yourself. This is the putting together of as much information as you can and organizing it under one unifying theme. After all, chemistry is moving so fast, and so much knowledge is accumulating, that you cannot afford merely to establish a mental card file of interesting but unconnected items.

You must, out of sheer efficiency, look for the broad fundamental concepts—learn them so well that you live with them in the lab and classroom—and then hang as much previously unrelated information on them as possible. As James B. Conant says, modern science is constantly trying to "lower the level of empiricism." You are a present-and-future-day scientist. I hope you are learning to think as one. I hope this book can serve as an example of how chemists use a great unifying concept.

There are some other attitudes about your work that I am trying to encourage, too. You will notice that I have tried to use illustrative problems which are real—not trumped up. Some of the data has come from laboratory note books. Most of it is from the chemical literature. The footnotes are there, not to make the book look erudite, but to suggest that you should begin to realize how important the scientific literature is to you. Get the habit of going to the library.

I have tried also to label the units and dimensions that belong with the numbers used to solve problems. This procedure goes by the high-sounding name of "quantity calculus." Don't worry about that—just see how helpful it is in keeping things straight. Your errors (and mine, most likely!) will be slips of the slide rule—not barkings-up-the-wrong-tree if you follow this method.

Many people helped me write this book—thousands of students, dozens of my former teachers, hundreds of friends now teaching you, and many whom I have met only through correspondence. I also should acknowledge with gratitude the patience of authors whose papers lay too long unattended on the desk of the Editor of the *Journal of Chemical Education* while this was being written.

Have fun with the mole—I did.

Sincerely yours,

WILLIAM F. KIEFFER

Wooster, Ohio
March, 1962

CONTENTS

An Open Letter	vii
1. Introduction	1
2. Gases	11
3. Stoichiometry: Formulas	22
4. Quantitative Relationships in Chemical Reactions	31
5. Properties of Liquid Solutions	47
6. Chemical Equivalence	63
7. Electrochemistry	83
8. Avogadro's Number	91
9. Calculation of Molecular Quantities	100
Summary	113
Index	115

chapter one

INTRODUCTION

The "mole" is a uniquely chemical concept. The modern chemist relies on it, even takes it for granted almost as he does such ideas as atom or molecule. The purpose of this book is to help students attain this automatic facility in making use of one of the most significant unifying concepts in all of general chemistry—the mole concept.

The word "mole" is derived from the Latin *moles*, meaning heap, mass, or pile. When we use the word "pile" to describe a collection of sand particles, peas, or ping pong balls, we have the closest analogy to the way the chemist uses the term "mole." The emphasis we will make repeatedly is that, regardless of how he defines *mole*, the chemist automatically thinks in terms of a particular *number* of particles. We will introduce the value of this number and our working definition of the term "mole" shortly.

First it will be helpful for us to realize how "automatic" much of a chemist's thinking is. For example, very early in his first course, the student learns that while the chemist may say in his mind, "Hydrogen gas reacts with oxygen gas to make water," he writes:

$$2H_2 + O_2 \rightarrow 2H_2O$$

2 the mole concept in chemistry

It is possible, if not probable, that this beginner may have lost points on his first quiz for writing:

$$H_2 + O \rightarrow H_2O$$

They were points well spent if he learned the crucial lesson that O and O_2 stand for completely different species. Consequently, the second equation describes a different reaction than does the first. Formulas identify ultimate kinds of particles called molecules. The adjective "ultimate" is important. It implies that the smallest possible number of individual chunks of matter to react as described above would be two pieces of hydrogen with one of oxygen. This is so logical that the student quickly "gets the hang of chemistry" and merrily balances all sorts of complicated equations by simply believing the formulas provided and counting the atoms to be sure none are lost.

Now our beginner is ready to encounter a sophisticated upperclassman who writes his equation:

$$H_2 + \tfrac{1}{2}O_2 \rightarrow H_2O$$

In the ensuing argument, we hope that our beginner's mind is flexible enough and our upperclassman eloquent enough to pave the way for full appreciation of what an equation means to a chemist. This last form of writing the equation is not a contradiction of the definition of molecule. Rather, it is an emphasis on what the coefficients in front of the formulas in an equation really mean: they are merely relative numbers. In the reaction of hydrogen gas with oxygen gas, only half as many molecules of the latter are involved. Said another equally valid way, twice as many hydrogens react as do oxygens. Our beginner can soon see ahead to predictions he can make from his equation. If he has some means of measuring numbers of hydrogen molecules (i.e., the amount of hydrogen gas), the coefficients in the balanced equation will tell him (without his actually having to measure it) the amount of oxygen which is required or the amount of water which is produced.

The big "if" in the last sentence is the important one for our beginner. When he goes into the laboratory, how can he count out a number of hydrogen molecules? What can a chemist measure which will tell him how many molecules he is working with? Suppose he did not know the formula of a substance, how could he find out what it ought to be? Are there some experiments or calculations in which it makes a difference whether he chooses 2 to 1 or 1 to $\frac{1}{2}$ as the ratio of coefficients? Are there some measurements which will have no relationship at all with the number of molecules reacting? After all, how is it that the chemist almost always makes his measurements macroscopically (in grams, pounds, tons, liters, cubic feet, etc.) yet he thinks microscopically by writing an equation as if he could pick out one or two tiny molecules[1] and have them react?

The more our beginner thinks about these questions, the more he realizes that when he has learned the answers to them he will begin to think as a chemist. These are the crucial questions; there must be adequate answers to them or there could be no modern chemistry or industry based upon it. Furthermore, as is typical with many of the most fundamental questions, there probably is some underlying concept which is so much an axiom for a chemist that he starts his thinking with it whenever a new problem arises. This axiom that the chemist takes for granted is the subject of this book.

Definitions

There are many ways of measuring the amount of a substance. Mass, weight and volume are some of the most com-

---

[1] It is pertinent to note the derivation of the term "molecule." Literally, it is the diminutive of mole and means *small* heap or pile. Modern connotation carries the implications of the ultimate or smallest heap, i.e., the single unit of which the mole is the larger "heap."

4 the mole concept in chemistry

mon. Concentrations (amount of one substance in a given amount of another) are measured by color, chemical reaction, taste, bacteriocidal effect, etc. The mole is the chemist's expression of amount. To him it means a definite number of individual unit particles of whatever species is being considered. This number is a standard. It is identified as Avogadro's Number. What it means we shall soon see. What its numerical value is, we need not know at once. We have then, the central theme of this book defined as:

The mole is Avogadro's Number of chemical units being considered.

The reader will note that this is as broad a definition as possible. In this discussion, a mole will be Avogadro's Number of atoms, molecules, ions, aggregates of ions, electrons or quanta—or of any other species the reader may choose.

This definition immediately demands another. What is Avogadro's Number? It may sound paradoxical, but we can tell what it is without knowing its numerical value—somewhat as we refer to the price of an article. If this kind of a definition of amount of substance is to have any utility, it must be related to other more simple means of measuring amount. Historically, the most obvious measure of amount was weight—the pull of gravity on the mass of an object. Accordingly, the first convenient measure of a mole was by weight.

As soon as the concept of atomic weights was established, it was recognized that if the weight of atom A was twice that of atom B, two pounds, grams, tons, or squeedunks of A would contain the same number of atoms as would one pound, gram, ton, or squeedunk of B. We can have some kind of Avogadro's Number (a constant number) of atoms in samples of two elements if the weights of the samples are in the ratio of the atomic weights of the elements. The magnitude of Avogadro's Number then depends upon two choices: the numerical scale of atomic weights and the weight units being used. The latter choice is the simpler; it can be established by decree. The former is much more elusive, since it is established by nature and can be discovered only by careful investigation and much research. After a true relative scale is uncovered, it can be made absolute by a decree establishing the choice of units.

Convention has established the gram as the unit of weight for molar amounts. This is merely a matter of consistency and convenience; in practice chemical engineers frequently use the pound-mole. If an element is assigned an atomic weight of x, *one mole* of that element will weigh x *grams*. Correspondingly, if a molecule has a molecular weight of xyz, *one mole* of the substance will weigh xyz *grams*. The perfectly general case is that based on a chemical formula (e.g., a salt in which there are no molecules). If the formula weight is pqr, *one mole* of the substance will weigh pqr *grams*.

Since the scale of atomic weights is so crucial to our definition, it is well for us to discuss the subject in some detail.

Atomic Weight Scales

The first scale of atomic weight values was that proposed by Dalton. Although his data were crude, his idea was sound. Hydrogen was assigned the value unity. Later the popularity of Prout's hypothesis that all atoms were compounded of the simplest atom, hydrogen, lent support to this. However, chemistry was not far into the 19th century when it became apparent that oxygen would be the best standard because it conveniently combines with many other elements. Berzelius in 1813 published a table which assigned oxygen the value 100.[2] He correctly assigned H = 6.64. (This can be translated into the modern value by taking the ratios: 6.64/100 = at. wt. H/16, from which we get: at. wt. H = 1.06). It is interesting to note that his values for Cu = 806, Hg = 2530, and several other

[2] See E. Farber, "The Evolution of Chemistry," the Ronald Press Company, New York, 1952, page 139.

metals reflected the typical dilemma of the time: the inability to decide how many atoms of each element were combined with one or more oxygen atoms. We will examine this problem further in Chapter 3.

One aid in the resolution of this question was the law reported by Du Long and Petit in 1819.[3] They made a discovery at the time thought to be empirical and almost unrelated to the problem: When the specific heat (the heat required to raise the temperature of a one gram sample one degree) of each of thirteen different metallic elements was multiplied by the corresponding atomic weight, a nearly constant number was obtained. This constant was 0.38 cal deg^{-1} on a scale that assigned oxygen an atomic weight of unity. By present standards this constant would have the value $0.38 \times 16 = 6.1$ cal deg^{-1}. The great advantage of such a discovery was that it provided an independent physical method for making a choice between various possible multiples of the oxygen-combining values. Thus, the Berzelius values for Cu and Hg had to be halved (806/2 = 403; 403/100 = at. wt. Cu/16; at. wt. Cu = 64.4 in reasonable agreement with the present value).

It is well to look at this law from a modern viewpoint. Probably without realizing it, Du Long and Petit had determined the first *molar* quantity. If we interpret their data in modern terms and let the atomic weight be the weight of a mole, we have in the case of their data for lead:

$$0.0293 \left(\frac{\text{cal}}{\text{g deg}}\right) \times 12.95 \times \left(\frac{16 \text{ g}}{\text{mole}}\right) = 6.07 \text{ cal deg}^{-1} \text{ mole}^{-1}$$

The temperature rise of a solid crystalline material upon the addition of heat is evidence of the increased vibrations of the atoms about their fixed positions in the solid. If the sizes of the samples of various metals are chosen so that the amount of heat

[3] See R. K. Fitzgerel and F. H. Verhoek, *Jour. Chem. Educ.* **37,** 545 (1960).

required to raise the temperature is the same (a constant 6 cal deg^{-1}), the suggestion follows that the numbers of vibrating atoms in the chosen samples are the same provided that the forces holding the atoms together are nearly the same strength. Modern theory corroborates this suggestion and predicts the magnitude of the constant amount of heat for a mole of metal atoms (6 cal deg^{-1}) from completely independent data; it also accounts for the apparent exceptions such as silicon and diamond which attain this value only at high temperatures.[4]

Not long after the work of Du Long and Petit, chemists began to agree that the value 16 for the atomic weight of oxygen would be the most convenient standard. Chemical techniques improved greatly, so that by the early decades of the present century, the atomic weights of elements were known with great accuracy. When Aston's pioneering work on mass spectroscopy made possible the direct evaluation of single isotopic masses, it was natural to refer all values of isotopic masses to the established oxygen scale of atomic weights. Gradually, the physical techniques improved so that isotopic abundance ratios, along with isotopic masses, made it possible to evaluate atomic weights for naturally occurring elements by the direct method of mass spectroscopy rather than to rely only on the relative method of measuring combining weight ratios.

In 1929, evidence from optical spectra showed that naturally occurring oxygen was a mixture of small amounts of ^{17}O and ^{18}O along with the abundant ^{16}O. This meant that since the precision of the mass spectrographic method was so high, it was necessary to call the mass of the ^{16}O isotope 16 exactly. Not only was the chemists' standard a mixture, but a mixture of slightly varying composition! This state of affairs has meant that since about 1940 there have existed *two* scales of atomic weights. The chemists' values based on natural oxygen = 16

[4] See Footnote 3, p. 6.

8 the mole concept in chemistry

exactly must be multiplied by the factor 1.000275 to agree with the physicists' values based on the scale $^{16}O = 16$ exactly.

Since the 1953 meeting of the Commission on Atomic Weights, International Union of Pure and Applied Chemistry, there has been considerable discussion of the possible adoption of a new unified standard scale. The chemists are loathe to change to the physicists' scale because at least millions of data on chemical systems, such as heats of reaction, would have to be changed in the literature. (The precision of these values exceeds the 3 parts in 10,000 that is represented by the conversion of one scale to the other). The physicists likewise object to changing to the chemists' scale for the obvious reason that "a spade should be called a spade." Various alternatives have been suggested, among them the use of the fluorine isotope 19 as the standard. This is virtually impossible for the mass spectrographer to do. He needs to use a nuclide of other than a prime number mass so that internal reference standards for the instrument can be conveniently established.

The nuclide chosen as the standard by international agreement in 1961 is the carbon isotope 12. We will not go into all the details of this fascinating chapter in the history of physical science.[5] The new scale will mean a revision of chemical data by a factor of only 43 parts per million downward. It will demand more of a revision in the physicists' values, but the advantages of ^{12}C as a primary standard in mass spectrometry outweigh this disadvantage.

The table that follows taken from the definitive article by Professor E. A. Guggenheim[6] summarizes the relationships between the various standards.

---

[5] The reader will profit greatly from reading the accounts of this problem and its solution. See, for example, *Jour. Chem. Educ.* **36**, 103 (1959) and the references there cited. Also E. Wichers, *Physics Today* **12**, No. 1, 28 (1959), and T. P. Kohman, J. H. E. Mattauch, and A. H. Wapstra, *Science* **127**, 1431 (1958).

[6] E. A. Guggenheim, *Jour. Chem. Educ.* **38**, 86 (1961).

	Old physical scale	Old chemical scale	New unified scale
^{16}O	16 exactly	15.99560	15.99491
O	.....	16 exactly	15.999
^{12}C	12.00382	12.00052	12 exactly
C	.....	12.011	12.010
Ag*	.....	107.873	107.868

*Value obtained from the recent coulometric determination at the National Bureau of Standards of the electrochemical equivalent of silver. W. R. Shields, D. N. Craig, and V. H. Dibeler, *J. Amer. Chem. Soc.* **82**, 5033 (1960).

The conversion factors between the scales are:

$$\frac{\text{Physical scale}}{\text{Chemical scale}} = \frac{16 \text{ exactly}}{15.99560} = 1.000275$$

$$\frac{\text{Physical scale}}{^{12}C \text{ scale}} = \frac{12.00382}{12 \text{ exactly}} = 1.000318$$

$$\frac{\text{Chemical scale}}{^{12}C \text{ scale}} = \frac{12.00052}{12 \text{ exactly}} = 1.000043$$

Summary and Redefinition

Now that agreement has been reached on all counts: the weight unit (grams), the scale of relative values (^{12}C scale) and what Avogadro's Number means, we can redefine the mole in very practical measurable terms as follows:[7]

The mole is the amount of substance containing the same number of chemical units as there are atoms in exactly 12 grams of ^{12}C.

The wording of this statement is in deliberate contrast to that on page 4. The mole now emerges as the fundamental quantity unequivocally established in operational terms. Avogadro's Number can now be defined with equal directness

Avogadro's Number, N, is the number of units in a mole

[7] See Footnote 6, p. 8.

10 the mole concept in chemistry

We are purposely avoiding the more popular terminology, "gram-atom," "gram molecular weight," "gram formula weight," etc., not because of semantic arguments but to make a simple emphasis.[8] The mole means more to the chemist than any of these or the "22.4 l (STP)" that is often its only meaning to a beginning student. We have defined (see p. 5) the mole in terms of a weight of substance. In Chapter 2 we will establish the complementary definition applicable to packages of molecules in the gaseous state. The measurements on gases are pressures, volumes and temperatures; the mole will be defined in terms of these units. In Chapter 7 we will be interested in the mole of electrons. There our concern will be not for their weight, nor volume but their electrical charge; the mole will be so defined (the faraday). In Chapter 9 we will recognize the mole of light quanta. The energy will be the chief concern; the mole will be so defined (the einstein).

---

[8] Readers who may wish to explore the matter of terminology further will be interested in the articles by G. N. Copley, S. Lee, and I. Cohen in *Jour. Chem. Educ.*, November, 1961.

chapter two

GASES

The connection between the properties of gases and numbers of molecules is as old as the idea of molecules itself. When Avogadro, in 1811, first proposed his hypothesis that equal volumes of gases contained equal numbers of molecules, he was providing an essentially chemical argument. He was explaining the unique whole number ratios that were exhibited by the volumes of reacting gases (Gay Lussac's Law). In the next chapter we will refer to the important application of Avogadro's ideas made by Cannizzaro to the problem of identifying molecular species and thence to establishing true relative atomic weights. This too was essentially a chemical argument.

Throughout the nineteenth century there also was developing what might be called a physical argument in support of Avogadro's brilliant guess. This was the kinetic molecular theory. Various kinds of information indicated that gases behave as they do because they are made up of molecules which move in random directions at great speeds and undergo almost elastic collisions.

The theory is so familiar that it seems almost an axiom to present day students. It describes a gas as being made up mostly of empty space. In fact, the "ideal" gas which the theory invents has a molecule with all its mass concentrated in a point. Yet the molecules have sufficient kinetic energy that their collisions with the walls of the container exert enough pressure to keep a volume available in which the molecules can move. Obviously, the pressure exerted by a gas will then depend on the number of molecules present to do the colliding. If an increase in temperature increases the kinetic energy of the molecules, each collision with the walls will exchange more momentum and the pressure will increase. This means that if pressure is to reflect the number of molecules in a gas sample, measurement must be made at a known temperature. It is also clear that even if temperature is constant, the same number of molecules will collide more often with the walls of a small volume than they will with the walls of a large volume.

Thus, we see that it should be possible to measure such variables as the pressure, volume, and temperature of a gas sample and have some reliable basis for indicating how many molecules it contains. In short, we have set up a system by which we can measure the *amount* of the gas. Common sense tells us that we also have the weight of a gas to tell us *amount*. Historically, the bringing together of these two approaches led to the development of the *mole* concept. It can be seen that a definite amount by weight of a specific material is one sure standard to establish. Then if this material can be made into a gas, the volume under agreed conditions can serve equally well, and sometimes more conveniently as a secondary standard for comparison.

It is important to emphasize the interrelationships of the various quantities in the establishment of standards. For almost a century chemists have agreed on assigning the convenient number 32 to the relative weight of an oxygen molecule.[1] If then, it is agreed to use an established standard unit of weight

---

[1]The problems in exact definition introduced by present-day knowledge about the variety of oxygen isotopes were discussed in Chapter 1. These are important both for precise statement and careful calculation to five significant figures. However, they need not complicate the story at this point.

(the gram), it is logically possible to translate relative weights of molecules into a real experimentally measurable set of quantities by choosing 32.00 grams as the weight of one mole of oxygen. Once the scale (O_2 = 32) and the unit (gram) are established, Avogadro's hypothesis (equal volumes contain the same number of molecules) backed up by the kinetic molecular theory, makes it possible to measure experimentally the weights of a mole of any other gas merely by weighing volumes of other gases.

> **EXAMPLE 2.1** The density of O_2 at 1 atm pressure, 0° C (standard temperature and pressure: STP) is 1.429 g l^{-1}. The density of helium is 0.1784 g l^{-1}; X gas is 1.163 g l^{-1}. What are weights of one mole of each gas?

We have been careful to state the question as "... the weight of one mole. ..." Numerically the answers are equal to the molecular weights. This, of course, is inherent in our initial choice of 32.00 grams as the weight of one mole of O_2. In practice, the chemist automatically identifies 32.00 with a "molecular weight." Hence, the answers are "molecular weights" to him.

Densities are given as the weight of a definite volume (g l^{-1}). Hence the answers are simply:

$$32.00 \left(\frac{g}{mole}\right)_{O_2} \times \frac{0.1784 \left(\frac{g}{l}\right)_{He}}{1.429 \left(\frac{g}{l}\right)_{O_2}} = 3.991 \text{ g mole}^{-1} = \text{weight of one mole of He}$$

$$\frac{32.00 \left(\frac{g}{mole}\right)}{1.429 \left(\frac{g}{liter}\right)_{O_2}} \times 1.163 \left(\frac{g}{l}\right)_X = 26.03 \text{ g mole}^{-1} = \text{weight of one mole of X}$$

It will be noted that the quantity 32.00/1.429 appears consistently in such problems. Indicating the appropriate division and dimensions further emphasizes its significance and usefulness:

$$\frac{32.00 \text{ g mole}^{-1}}{1.429 \text{ g l}^{-1}} = 22.39 \text{ liters mole}^{-1} \text{ (STP)}$$

This figure is the volume (STP) which is occupied by a mole of O_2 gas. A mole of any other gas should also occupy the same volume because the number of molecules will be the same. Note that the actual number of molecules is still unknown. An entirely different line of attack is needed to count the molecules —just as it would require a completely different kind of experiment to weigh a single molecule. Even though the number of molecules in a mole is unknown, it can be seen from the arguments thus far that this number is one of the most relied upon constants of nature. Our postponing both a discussion of how it is evaluated and our giving the actual figure is calculated further to emphasize that chemists can do a great deal of their work knowing only that there is such an unchanging number (see Chapter 8).

The student who has already been introduced to chemistry may have thought that the number 22.39 above contained a misprint. Probably other books have told him that a mole of O_2 occupies 22.414 l (STP). Certainly the table of atomic weights tells him that helium has a molecular weight of 4.003. If we are being careful both about significant figures in our calculations and about accuracy in definition, what is wrong?

The difficulty, of course, is not with our arithmetic, but with the experiment we have chosen as a basis for definition. As soon as we are willing to make precise measurements on gases, we must recognize that our expectation that all molecules of gases will behave the same regardless of their identity (Avogadro's hypothesis and the kinetic molecular theory) is bound to be only a first approximation. This means that we must examine gas behavior and possibly be forced to abandon real gases in favor of a hypothetical, well-regulated "ideal gas" to give a foundation for definition.

The combined laws of Boyle and Charles predict that the quantity $(P \times V)/T$ = constant for any particular sample of any particular gas. It can also be seen that if P and T are unchanged, an increase in the amount (weight) of gas will increase the volume. The equation can then be written: $(P \times V)/T = \text{(constant)}' \times g$. If the weight g is translated into number of moles $n = g/\text{g mole}^{-1}$, then the equation will read:

$$\frac{P \times V}{T} = nR$$

The symbol "R" usually represents the constant when the equation is written with n. Note that the equation specifies much about the dimensions of R.[2]

$$R = \frac{\text{pressure} \times \text{volume}}{\text{number of moles} \times \text{absolute temp.}}$$

One way of testing the behavior of gases to see how well they follow Boyle's and Charles' laws would be to observe how the product of $P \times V$ for constant n varies as the pressure is increased. As the pressure increases, the volume should decrease proportionally. In the case of helium, the volume does not decrease as much as expected, so the quantity PV/nT increases slightly as the pressure is increased. Oxygen, in common with most gases, behaves differently. As the pressure is increased

[2] Further examination of the $P \times V$ part of this equation reveals that since $P = \text{force/distance}^2$ and $V = \text{distance}^3$, the product $PV = \text{force} \times \text{distance} = \text{energy}$. Hence any correct value of R must be expressed as energy mole^{-1} (°K)$^{-1}$. For example, R has the value 8.314×10^7 erg mole^{-1} (°K)$^{-1}$ or 1.987 calories mole^{-1} (°K)$^{-1}$.

to 100 atm, the value of PV/nT steadily decreases. Data on many gases at a variety of temperatures indicate what one would expect from a reexamination of the postulates of the kinetic molecular theory. Real gases will follow the gas laws only when the pressure is vanishingly small. Another way of saying it is that as $P \rightarrow 0$, all gases will exhibit ideal gas behavior:

$$\frac{P \times V}{T} = R \times n$$

This is called the *equation of state* for an ideal gas. The limiting value of PV/T as $P \rightarrow 0$ turns out to be 0.082054 l atm mole^{-1} ($^\circ$K)$^{-1}$. Thus, we have a value for R which correctly represents the proportionality constant between the behavior of an ideal gas (numbers for the variables P, V, and T) and the number of moles of gas in the sample.

$$R = 0.082054 \text{ l atm mole}^{-1} (^\circ\text{K})^{-1}$$

If this value for R is used to calculate *the volume for one mole of an ideal gas at $0^\circ C$ ($273.16\,^\circ K$) at one atmosphere pressure*, we have

$$V = \frac{nRT}{P} = \frac{1.0000 \text{ (mole)} \times 0.082054 \left(\frac{\text{l atm}}{\text{mole }^\circ\text{K}}\right)}{1.0000 \text{ (atm)}} \times 273.16 \text{ (}^\circ\text{K)}$$

$V = 22.414$ l (STP) for one mole of an ideal gas.

The number 22.414 l (STP) is probably one of the best remembered numbers in general chemistry. We hope that readers will realize that it is the STP volume for a mole of *ideal gas*. Real gases only approximate it. Even oxygen, the long established weight standard for defining a mole, does not actually have a molar volume of 22.414 liters (STP). We have already shown it experimentally to be 22.39 liters. A few other experimentally measured molar volumes (STP) are: helium, 22.426 l; methane, 22.360 l; hydrogen bromide, 22.212 l; ammonia, 22.09 l.

The variation in these numbers is an indication of the degree to which Avogadro's hypothesis is only approximately true. In the case of ammonia the deviation is about 1.4%. Usually the values of the molar volume are lower than the ideal because of the contracting forces of intermolecular attraction. Helium (and H_2) are exceptions, being less compressible than an ideal gas at $0°C$. The reader will now be able to account for the puzzling answers to Example 2.1. The comparison there was made between O_2, a gas which deviates in a negative direction from the ideal gas molar volume, and He, a gas which has a small positive deviation. If the comparison were made between the actual behavior of He, a real gas, and the hypothetical ideal gas, the molecular weight of He would be very close to the expected theoretical value.

EXAMPLE 2.2 The measured density (STP) of He is 0.1784 $g\ l^{-1}$. What is the weight of one mole?

The weight of one mole will be the weight of 22.414 l (STP). Since the density is the weight of one liter, we have simply:

$$\text{weight of one mole} = 0.1784 \left(\frac{g}{l}\right) \times 22.414 \left(\frac{l}{\text{mole}}\right) = 4.000\ g\ \text{mole}^{-1}$$

The difference between this value and the "true" molecular weight now reflects only the degree to which He deviates from ideal gas behavior.

The reader who is looking for broadly applicable generalizations (and every beginning science student should!) has probably already framed one from the discussion so far. It is:

$$\text{weight of one mole} = \text{STP density} \left(\frac{g}{l}\right) \times 22.414\ \text{liters}$$

This is true. Moreover, very often the chemist is faced with just this problem of evaluating the weight of one mole of a gas. The above generalization would suggest that the data needed are the STP density. Even if density data were obtained for the gas at some other set of pressure and temperature conditions, the STP density can be calculated by applying Boyle's and Charles' laws. This type of calculation is discussed in every general chemistry textbook and need not be elaborated here.

We ask the reader to recognize that his essential problem is the calculation of the number of moles of material. We have demonstrated thus far in this chapter that this quantity can be obtained for gases by applying the equation of state (the expected relationship between the variables P, V and T) for an ideal gas. We have then:

$$\text{number of moles} = n = \frac{P\,(\text{atm}) \times V\,(\text{l})}{R\,(1\text{ atm mole}^{-1}(°\text{K})^{-1}) \times T\,(°\text{K})}$$

Note that attention always must be given to the units in which the variables are expressed so that a correct choice for R can be made. We shall consistently use those chosen above (atm, l, °K and g) only because of convenience.

> **EXAMPLE 2.3** Show that R can be expressed as:
>
> $$R = 82.05 \text{ ml atm mole}^{-1}\,(°\text{K})^{-1}$$
>
> or $62{,}360$ ml mm Hg mole^{-1} (°K)$^{-1}$

This will be left as an exercise for the reader.

Engineering data often use the English system of units. The use of pounds weight suggests a pound mole rather than a (gram) mole. Temperatures in Fahrenheit must be converted to degrees Rankine, the "absolute" Fahrenheit (°R = 459.7 + °F). Pressures can be lb in.$^{-2}$ and volumes in ft^3. With these units, the value of R is

$$10.73 \text{ ft}^3\,(\text{lb in.}^{-2})\,(\text{lb mole})^{-1}\,(°\text{R})^{-1}$$

Confirming this is a good exercise for the reader.

A typical method for evaluating the weight of a mole of gas is the Dumas procedure. A glass bulb of known weight and volume is filled with vapor of an unknown substance. This is most easily accomplished if the substance is added as a liquid, vaporized at a known temperature and pressure, the bulb sealed, and weighed.

> **EXAMPLE 2.4** A 0.501 g sample filled a bulb of 203 ml volume at 99° C, 735 mm pressure. What is the weight of a mole?

The number of moles can be represented in two ways:

$$\frac{\text{weight}}{\text{weight of one mole}} = n = \frac{P \times V}{R \times T}$$

The numerical value for the number of moles need not be found, since the equality is used to calculate the weight of one mole:

$$\frac{0.501 \text{ (g)}}{M \left(\frac{\text{g}}{\text{mole}}\right)} = \frac{\frac{735}{760} \text{(atm)} \times 0.203 \text{ (l)}}{0.0820 \left(\frac{\text{l atm}}{\text{mole}^\circ \text{K}}\right) \times 372 \text{ (°K)}}$$

$$M = \frac{0.501 \times 0.0820 \times 372 \times 760}{735 \times 0.203} = 77.9 \text{ g mole}^{-1}$$

Let us contrast this method with one implied by remembering the "22.4, (STP)" number: First, calculate the STP volume for the sample:

$$V \text{ STP} = 0.203 \text{ (l)} \times \frac{735 \text{ (mm Hg)}}{760 \text{ (mm Hg)}} \times \frac{273 \text{ (°K)}}{372 \text{ (°K)}}$$

Then divide by 22.4 to get the number of moles:

$$n = 0.203 \text{ (l)} \times \frac{735 \text{ (mm Hg)}}{760 \text{ (mm Hg)}} \times \frac{273 \text{ (°K)}}{372 \text{ (°K)}} \times \frac{1}{22.4 \text{ (l)}}$$

20 the mole concept in chemistry

This approach uses two steps and two arbitrary numbers which must be remembered (273 and 22.4) in lieu of the single step and one number (0.0820). More important than the efficiency, however, is the emphasis on direct calculation of the chemist's most useful conceptual quantity, *number of moles*.

Another typical general chemistry problem is the conversion of student laboratory data into "chemical" information. The following is illustrative:

> EXAMPLE 2.5 A student collected 51.2 ml of H_2 gas over water at 18° C (vapor pressure = 15.5 mm Hg) at a barometric pressure of 736 mm Hg. What fraction of a mole of gas is in his sample?

$$n = \frac{\frac{736 - 15}{760} \text{ (atm)} \times 0.0512 \text{ (l)}}{0.0820 \left(\frac{1 \text{ atm}}{\text{mole °K}}\right) \times 291 \text{ (°K)}} = 0.00203 \text{ mole}$$

The composition of mixtures of gases usually is expressed in percent by volume. The chemist, used to thinking automatically in terms of Avogadro's hypothesis, immediately translates this into mole-percent or mole-fraction. Thus, if we have a mixture made up of gases A, B, C, we have

$$\text{volume \% of gas A} = \frac{\text{volume of gas A}}{\text{total volume of gas (A + B + C) mixture}} \times 100$$

$$\text{mole fraction of gas A} = \frac{\text{number of moles of gas A}}{\text{total number of moles of gas (A + B + C) mixture}}$$

and similarly for B and C. Often a gas mixture can be analyzed at constant volume by removing one or more constituents. This can be done either by freezing the constituent out of the sample or by reacting it with some non-volatile reagent. The drop in pressure is proportional to the number of moles removed. The following is an example.

EXAMPLE 2.6 A mixture of H_2O vapor, CO_2 and N_2, was trapped in a glass apparatus with a volume of 0.731 ml. The pressure of the total mixture was 1.74 mm Hg at 23° C. The sample was transferred to a bulb in contact with dry ice ($-75°$ C) so that H_2O vapor was frozen out. When the sample was returned to the measured volume, the pressure was 1.32 mm Hg. The sample was then transferred to a bulb in contact with liquid nitrogen ($-195°$ C) to freeze out the CO_2. In the measured volume, the pressure was 0.53 mm Hg. How many moles of each constituent are in the mixture?

The temperature of the measured volume is constant, so the partial pressures of each constituent will reflect accurately the relative number of moles. The composition then is:

$$\text{mole fraction of } H_2O = \frac{1.74 - 1.32}{1.74} = \frac{0.42}{1.74} = 0.24$$

$$\text{mole fraction of } CO_2 = \frac{1.32 - 0.53}{1.74} = \frac{0.79}{1.74} = 0.45$$

$$\text{mole fraction of } N_2 = \frac{0.53}{1.74} = 0.31$$

The total number of moles is:

$$n = \frac{P \times V}{R \times T} = \frac{1.74 \,(\text{mm Hg}) \times 0.731 \,(\text{ml})}{62,360 \left(\frac{\text{mm Hg ml}}{\text{mole}\,°K}\right) \times 296 \,(°K)} = 6.9 \times 10^{-8} \text{ mole}$$

The number of moles of H_2O in the mixture is $0.24 \times 6.9 \times 10^{-8} = 1.7 \times 10^{-8}$. Correspondingly, $CO_2 = 3.1 \times 10^{-8}$ mole; $N_2 = 2.1 \times 10^{-8}$ mole.

chapter three

STOICHIOMETRY: FORMULAS

Stoichiometry is the single word (derived from the Greek *stoicheion*, meaning element) which chemists use broadly to describe the science of chemical arithmetic. In a strict sense, the term applies to calculations based upon the ratio of the masses of the elements which have combined to form a pure compound. An obvious extension includes the calculations based on the ratios of the masses of all substances (elements or compounds) which react according to a known chemical equation. The reader immediately recognizes that the word "ratio" reminds the chemist that atomic weights can serve as his starting point along with the appropriate combinations of atomic weights into formula weights. Furthermore, the working chemist realizes that stoichiometry includes, for him, what might loosely be called a "philosophy" of reasoning based on chemical calculation. What we hope to do in this and subsequent chapters is to spell out, at the risk of being obvious, how the mole concept is the basis of this approach.

It is hard for present-day chemistry students to realize what chaos the science was in during the first half of the nineteenth century. For over fifty years after Dalton had established the idea of atomic weights, chemists were arguing about them. Everyone realized that the best they could do was to establish relative weights. Likewise, they knew it was hopeless for them ever to expect to weigh a single minute atom. Their difficulty came from their never being able to decide whether they were weighing equal or multiple numbers of atoms. A moment's reflection shows that there is no sure way out of this dilemma if the only data at hand are the combining weights of elements.

Imagine trying to decide from the single fact that 1.0 weight of hydrogen combines with 8.0 weights of oxygen, whether the formula should be HO or H_2O. If the atomic weight of H = 1 and O = 8, the former is correct; if the choice is made that H = 1 and O = 16, the second must be true. The number of possibilities depends entirely upon the number of choices for the weights of H and O atoms. (In passing, we may note that the problem was complicated by Dalton's preference for HO).

We cannot stop here to recount the story of the famous Karlsruhe Congress of 1860[1] which was held in the hope that some order could be made out of the conflicting claims by different interpreters for the same data. The majority of those who came to this first international conclave of chemists saw no way out of the dilemma imposed by working only with gravimetric data. Many went home convinced that a paper by Cannizzaro held the key. He had shown that Avogadro's hypothesis, enunciated nearly 50 years before gave chemists a completely independent yet equally valid line of attack on establishing the identity of molecules and measuring their relative weights. (If equal volumes of gases under identical conditions of pressure and temperature contain equal numbers of molecules, then the weights of these equal volumes have to be in the ratio of the weights of the individual molecules of the gases.) Once these molecular weights were known, the gravimetric analysis information could reveal a characteristic weight or multiple of a characteristic weight for each element in as many of its compounds as could be vaporized.

A present-day student can read this chemical history with in-

[1] See, for example, A. J. Ihde, *Jour. Chem. Educ.* **38,** 83 (1961).

24 the mole concept in chemistry

terest and appreciation after he has repeated the type of calculations that are involved. Most textbooks of general chemistry provide this kind of exercise. It is hardly pertinent, however, for our treatment here to repeat the calculations of research done years ago which have provided the data in the table of atomic weights. Rather we choose problems illustrative of those which are the kind modern chemists may encounter and solve with available resources of standard tabular data.

Establishing Simplest Formulas

The first of these is to use the accepted values of atomic weights and gravimetric data to reveal the correct numbers of atoms or aggregates of atoms in a given compound.

> **EXAMPLE 3.1** 3.75 g of aluminum was found to produce 37.06 g of a compound with bromine. What is the simplest formula for the compound?

What we want to know is n and m in the formula $Al_n Br_m$. These are the number of atoms (or, at the same time the number of moles) of each element in combination.

In this and in all cases hereafter in this discussion, where the weight of one mole of an atom, molecule or unit designated by a formula is involved, we will take it for granted that the reader has available a table of atomic weights for reference. (See page facing inside rear cover.) Thus the number of moles of any substance will be $g/g\ mole^{-1}$ as indicated by the symbols.

$$n = \text{moles of Al} = \frac{3.75\ (g)}{26.98\ \left(\frac{g}{mole}\right)} = 0.139\ \text{mole}$$

$$m = \text{moles of Br} = \frac{37.06 - 3.75\ (g)}{79.92\ \left(\frac{g}{mole}\right)} = 0.417\ \text{mole}$$

$Al_{0.139}Br_{0.417}$ is a true but inconvenient formula. Hence convention suggests:

$$Al_{\frac{0.139}{0.139}}Br_{\frac{0.417}{0.139}} = AlBr_3.$$

Notice that if we follow strictly our vocabulary which specifies a mole on the basis of grams and the atomic weight scale, calling n the *number of moles* of Al would mean that if we were given 3.75 pounds of Al, we would have to introduce the conversion factor 453.6 grams pound^{-1}. However, the same would be true if 33.31 pounds of bromine were involved so that the conversion factors would cancel in the final step.

The most frequent calculation of this type made by chemists is the confirmation of a formula implied by other lines of research. Usually the claims of a chemist that he has synthesized a new compound are not considered valid unless they include the critical test of having the elemental analysis agree with that predicted by the formula.

> **EXAMPLE 3.2** A whole new branch of chemistry opened up in 1952 with the announcement of the synthesis of a carbon, hydrogen, and iron compound eventually called "Ferrocene." One of the initial announcements was made by Miller, Tebboth, and Tremaine.[2] They reported C = 64.4%, H = 5.5%, Fe = 29.9%. Show why they concluded that the simplest formula for the compound was $C_{10}H_{10}Fe$.

Note that analytical data given as percentages are already conveniently *relative*, so that numbers of grams equal to the percentages may be considered for the purists who want to define the following subscripts strictly as numbers of moles.

[2] S. A. Miller, J. A. Tebboth, and J. F. Tremaine, *Jour. Chem. Soc.* (London) **1952**, 632.

$$\frac{C_{64.4\,(g)}}{12.0\left(\frac{g}{mole}\right)} \quad \frac{H_{5.5\,(g)}}{1.00\left(\frac{g}{mole}\right)} \quad \frac{Fe_{29.9\,(g)}}{55.8\left(\frac{g}{mole}\right)}$$

$$C_{5.37}H_{5.5}Fe_{0.536}$$

EXAMPLE 3.3 Any copy of the *Journal of the American Chemical Society* at random will supply numerous illustrations of this type of problem. What formula do the authors claim for the carbon-hydrogen-oxygen compound which yielded C = 86.47%, H = 6.43%? (The remaining percentage is oxygen.) Answer: $C_{16}H_{14}O$. See *Jour. Am. Chem. Soc.* **83**, 2142 (1961).

An extension to this type of calculation can be applied to such compounds as hydrates or those which contain co-ordinated ammonia.

EXAMPLE 3.4 A student heated 1.763 g of hydrated $BaCl_2$ to dryness. 1.505 g of the anhydrous salt remained. What is the formula of the hydrate?

The problem is to evaluate the unknown number of moles of H_2O associated with one mole of $BaCl_2$. In order to write a formula, we want to find m and n in:

$$m\ BaCl_2 \cdot n\ H_2O$$

$$\frac{1.505\,(g)}{208.3\left(\frac{g}{mole}\right)} BaCl_2 \cdot \frac{(1.763 - 1.505)\,(g)}{18.0\left(\frac{g}{mole}\right)} H_2O$$

$$0.00722\ (mole)\ BaCl_2 \cdot 0.0143\ (mole)\ H_2O$$

$$BaCl_2 \cdot 2\ H_2O$$

> **EXAMPLE 3.5** Often in the older literature of mineralogy the composition of claylike materials was given in terms of the oxides. For example, the mineral *talc* was expressed as MgO = 31.88%, SiO$_2$ = 63.37%, and H$_2$O = 4.75%. Show that a simplest formula for talc is H$_2$Mg$_3$Si$_4$O$_{12}$. (Hint: adapt the procedure used in Example 3.4 and then rearrange the symbols.)

This will be left as an exercise for the reader.

Non-Stoichiometric Compounds

Modern techniques of examining materials in the solid state have revived a concept that was first proposed by Berthollet in 1803 at about the time Dalton proposed his atomic theory. Berthollet erroneously claimed that the composition of certain compounds was variable, in contradiction to the Law of Definite Proportions. At that time the best analysts of the day were able to refute Berthollet's claims and substantiate Dalton's. However, it is now known that many crystalline inorganic solids can accommodate varying numbers of atoms of one element relative to the number of atoms of another element. The commonest examples are among the metallic oxides and sulfides. In honor of the great Berthollet whose imagination exceeded his manipulative abilities these are called *berthollide* compounds.

One interesting example of a berthollide is ZnO. When the white crystals of pure ZnO (atomic ratio 1/1) are heated in zinc vapor, the crystal assumes a brilliant reddish orange color. Analysis indicates that on a weight basis it has taken up as much as 0.03% excess zinc. These additional zinc atoms find interstitial sites in the crystal which can accommodate them. When one realizes how many atoms are involved in any actually measurable sample of a solid crystal, the possibility of berthollide compound existence seems more reasonable. For example, in the case of ZnO, since the atomic weight of Zn is 65.38 compared to oxygen's 16.00, then translating the 0.03 weight % of excess zinc into atom fraction the result would be only $(16/65.38) \times 0.0003 = 0.00007$. This is the fraction by which the number of Zn atoms exceeds the number of O atoms. (The value of Avogadro's Number, the actual number of atoms in one mole, is 6.023×10^{23}. See Chapter 8.) So, the factor 0.00007 would mean that about 4.2×10^{19} extra Zn atoms would be present in one mole of crystalline ZnO. If we were to write the "correct" formula, it would be

$$Zn_{1.00007}O_{1.00000} \text{ or } Zn_{6.023 \times 10^{23} + 4.2 \times 10^{19}}O_{6.023 \times 10^{23}}.$$

Other berthollides vary over much wider ranges of composition. A sulfide of cerium appears to be able to accomodate a great many additional cerium atoms in its crystal lattice. Brewer[3] describes what happens as cerium is added to Ce_2S_3 before a sufficient amount is present to establish the "true" compound Ce_3S_4.

EXAMPLE 3.6 If analysis demonstrated that a cerium sulfide contained 74.7% cerium, what formula would have to be assigned the material? *Answer:* $Ce_{2.7}S_4$.

Establishing Molecular Formulas

In the previous chapter we showed how the measurement which can be made on a sample of material in the gaseous state can be translated into number of moles or to establish the weight of one mole. For a gas containing molecules as the particular species, this technique gives a true "molecular weight" (the number of grams in a mole of molecules). This is im-

[3] L. Brewer, *Jour. Chem. Educ.* **38**, 91 (1961).

portant information to have, if the decision is to be made on the question of whether the simplest formula is the true molecular formula. Consider, for example, the substance for which data were given in Example 2.4. The weight of a mole of this vapor was 77.9 g. Even if we know that the only elements in it are carbon and hydrogen, we still cannot establish its formula. If, however, we know the composition in terms of weight percent, we can learn the formula as follows:

EXAMPLE 3.7 One mole of a gas weighs 77.9 g; analysis reveals that it is 92.4% carbon, the remainder, hydrogen. What is its molecular formula?

We have again the problem of establishing n and m in the formula: C_nH_m. However, we can proceed directly to a specific value of n and m rather than *relative* values as was our limit in Examples 3.1–3.3. The weight of one mole of the molecule C_nH_m must be the aggregate of the weights of n moles of C atoms and m moles of H atoms. Thus:

$$\text{weight of } n \text{ moles of C in one mole of } C_nH_m = 0.924 \times 77.9 \ \left(\frac{g}{\text{mole}}\right)$$

$$\text{weight of } n \text{ moles of C} = n \text{ (mole)} \times 12.0 \ \left(\frac{g}{\text{mole}}\right)$$

$$n = \frac{0.924 \times 77.9 \ (g)}{12.0 \ \left(\frac{g}{\text{mole}}\right)} = 6 \text{ moles of C atoms.}$$

$$\text{Similarly } m = \frac{0.076 \times 77.9 \ (g)}{1.01 \ \left(\frac{g}{\text{mole}}\right)} = 6 \text{ moles of H atoms.}$$

$$\text{formula} = C_6H_6$$

30 the mole concept in chemistry

> **EXAMPLE 3.8** A 0.0553 g sample of a boron-hydrogen compound created a pressure of 0.658 atm in a bulb of 40.7 ml volume at 100° C. Analysis showed it to be 85.7% boron. Show that the molecular formula is B_5H_9.

This problem merely combines features of previous ones. The first step is to equate the number of moles of sample calculated by two approaches. From the resulting equation the weight of one mole can be found. Thence, we have the molecular formula

$$\frac{0.0553_{(g)}}{M\left(\frac{g}{mole}\right)} = \begin{array}{c}\text{number}\\\text{of}\\\text{moles}\end{array} = \frac{0.658_{(atm)} \times 0.0407_{(l)}}{0.0820\left(\frac{l\,atm}{mole\,°K}\right) \times 373_{(°K)}} = 0.000875 \text{ mole}$$

$$M = 63.2 \text{ g mole}^{-1}$$

$$B_nH_m = B_{\underset{10.8\left(\frac{g}{mole}\right)}{0.857 \times 63.2_{(g)}}} H_{\underset{1.01\left(\frac{g}{mole}\right)}{0.143 \times 63.2_{(g)}}} = B_{5.02}H_{8.96}$$

chapter four

QUANTITATIVE RELATIONSHIPS IN CHEMICAL REACTIONS

The chemist uses his shorthand to give him a maximum of information. The familiar chemical equation is an elegant example. Whether it is the simple: $2Ca + O_2 \rightarrow 2CaO$ or the apparently more complicated: $H_2PO_4^- + UO_2(C_2H_3O_2)_2 \rightarrow HC_2H_3O_2 + C_2H_3O_2^- + UO_2HPO_4$, the chemist knows that he will be able to interpret accurately any macroscopic observations (such as volumes of gases or weights of solids) in terms of the microscopic entities, atoms, molecules, ions, etc. This is because his automatic thinking translates the coefficients in front of the formulas into relative numbers of moles of reacting species. Thus he immediately recognizes that

$$2\,Ca + O_2 \rightarrow 2\,CaO$$
$$2 \text{ mole} + 1 \text{ mole} \rightarrow 2 \text{ mole}$$
$$\text{or } 1 \text{ mole} + \tfrac{1}{2} \text{ mole} \rightarrow 1 \text{ mole}$$
$$\text{or umpteen mole} + \frac{\text{umpteen}}{2} \text{ mole} \rightarrow \text{umpteen mole}$$

In passing we should note that although the law of conservation of mass, expressed in the concepts of atomic theory, means

32 the mole concept in chemistry

a law of "conservation of numbers of atoms," there is no such thing as a law of "conservation of formula moles." The *three* moles of reactants expressed by the formulas $2Ca + 1O_2$ make *two* moles of products expressed by the formula $2CaO$.

Almost always the chemist's problem will simply be calculating the value of "umpteen" in the last equality. One of these "umpteens" to serve as a starting point will be revealed by his experimental data:

$$\text{number of moles} = \frac{\text{weight of substance (g)}}{\text{weight of one mole of substance}\left(\frac{g}{mole}\right)}$$

He recognizes that the coefficients of the balanced equation let him know at a glance how many umpteen moles of any other substance will be involved. Usually his data for solids will be weights, although later we will see that often it is more convenient to put materials into solvents and measure volumes of liquid solutions.

> **EXAMPLE 4.1** What weight of oxygen will react with 1.00 g of calcium?

The balanced equation shows that the ratio of oxygen to calcium required theoretically is:

$$\frac{\text{oxygen}}{\text{calcium}} = \frac{O_2}{2Ca} = \frac{1 \text{ mole}}{2 \text{ mole}} = \frac{0.5 \text{ mole}}{1.0 \text{ mole}}$$

Then the specific data of the problem fit directly into the same sequence of ratios:

$$\frac{\text{moles of oxygen required}}{\text{moles of calcium given}} = \frac{x \text{ (mole) of } O_2}{\dfrac{1.00 \text{ (g)}}{40.1 \left(\frac{g}{mole}\right)} \text{ of Ca}} = \frac{x \text{ (mole) of } O_2}{0.025 \text{ (mole) of Ca}}$$

quantitative relationships in chemical reactions

The solution to the problem depends on equating the demands of the "theoretical" equation with the problem specifications.

$$\frac{\text{oxygen}}{\text{calcium}} = \underbrace{\frac{0.5 \text{ (mole)}_{O_2}}{1.0 \text{ (mole)}_{Ca}}}_{\substack{\text{equation} \\ \text{prediction}}} = \underbrace{\frac{x \text{ (mole)}_{O_2}}{0.025 \text{ (mole)}_{Ca}}}_{\substack{\text{experimental} \\ \text{data}}}$$

The number of moles of O_2 required to react with 1.00 g of Ca is then $0.5 \times 0.025 = 0.0125$ mole. This can be translated into grams simply by multiplying by the unit weight of a mole (in grams) of oxygen:

$$\text{oxygen required} = 0.0125 \text{ (mole)} \times 32.0 \left(\frac{\text{g}}{\text{mole}}\right) = 0.400 \text{ g}$$

The student who may be used to plugging members into an "is-to-" proportion may feel that we have gone "round Robin Hood's barn" to get the answer he gets simply by:

$$\begin{array}{cc} 1.00 & y \\ 2\text{Ca} & + \quad O_2 \\ 2 \times 40.1 & 32.0 \end{array} \qquad y = \frac{32.0 \times 1.00}{2 \times 40.1} = 0.400$$

This is true, as must be the case with all empirical methods which produce correct answers. However, such a student owes himself the full understanding of why he sets up the numbers he does. Examination will reveal that he has performed exactly the same steps.

$$y \text{ g of oxygen} = \underbrace{32.0}_{\substack{\text{weight} \\ \text{of one} \\ \text{mole of } O_2}} \times \underbrace{\frac{1}{2}}_{\substack{\text{ratio:} \\ \text{moles of } O_2 \\ \text{moles of Ca} \\ \text{established} \\ \text{by equation}}} \times \underbrace{\frac{1.00}{40.1}}_{\substack{\text{number of} \\ \text{moles of Ca} \\ \text{set by} \\ \text{the} \\ \text{problem}}}$$

or in completely general form:

weight of product = weight of one mole of product × ratio from equation: $\dfrac{\text{moles of product}}{\text{moles of reactant}}$ × number of moles of reactant given

> **EXAMPLE 4.2** What weight of CO is required to form $Re_2(CO)_{10}$ from 2.50 g of Re_2O_7 according to the equation:
> $$Re_2O_7 + 17CO \rightarrow Re_2(CO)_{10} + 7CO_2$$

$$y \text{ g of CO} = \left[28.0 \left(\frac{g}{mole}\right)_{CO}\right] \times \left[\frac{17 \text{ (mole)}_{CO}}{1 \text{ (mole)}_{Re_2O_7}}\right] \times \left[\frac{2.50 \text{ (g)}_{Re_2O_7}}{484.4 \left(\frac{g}{mole}\right)_{Re_2O_7}}\right]$$

$$y = 2.46 \text{ g of CO}$$

The reader should convince himself of the significance of each bracketed term as being illustrative of the general relationship. The example is chosen not for its importance, but to demonstrate the simplicity and universal applicability of the approach. Countless other examples could be given. It is equally clear that if the above problem had read pounds or tons of Re_2O_7, the answer would appear in the same units as those chosen for the given weight. The third bracket, (2.50/484.4) would, of course, have to be read as number of "pound-moles," "ton-moles" or whatever unit was involved.

There are times when it is much more convenient to measure the amount of a material by other than direct weighing. In the following example, a volume of reagent is specified. It must be realized that this requires that density data be provided.

EXAMPLE 4.3 28.5 ml of benzene (C_6H_6, density = 0.88 g ml^{-1}) is reacted with excess Br_2 to produce the compound C_6H_5Br. What is the maximum weight of compound to be expected if the reaction is simply: $C_6H_6 + Br_2 \rightarrow C_6H_5Br + HBr$.

$$\text{weight of } C_6H_5Br = \left[157 \left(\frac{g}{mole}\right)_{C_6H_5Br}\right] \times \left[\frac{1 \text{ (mole)}_{C_6H_5Br}}{1 \text{ (mole)}_{C_6H_6}}\right] \times \underbrace{\left[\frac{28.5 \text{ (ml)}_{C_6H_6} \times 0.88 \left(\frac{g}{ml}\right)_{C_6H_6}}{78.1 \left(\frac{g}{mole}\right)_{C_6H_6}}\right]}_{\text{number of moles of } C_6H_6 \text{ in 28.5 ml}}$$

$$y = 50.4 \text{ g of } C_6H_5Br$$

Gravimetric Factors

There are many cases in which a particular chemical reaction serves as the basis for repeated analyses of a routine nature. For example, it may be necessary to monitor the control of an industrial process so that uniformity of product can be assured. Or, it may be desirable to label each batch of product with the carefully determined amounts of minor impurities. There are times, too, when it is much more convenient to perform an analysis to get one material, and to translate the amount of it into figures representing something else. For example, in the analysis of an ore, it is usually much easier to isolate a compound of a metal such as an oxide than to obtain the pure metal itself. If then the "theoretical" amount of pure metal in the compound is known, a multiplicative factor will translate weight of compound into weight of metal.

Whenever an analysis is routine, the chemist will combine as much of the necessary arithmetic as possible into a single number which can serve as a constant factor in his calculation. The principle involved can easily be recognized from a reexamination of our generalization; (page 34):

$$\frac{\text{wt. of}}{\text{product}} = \frac{\text{wt. of one mole}}{\text{of product}} \times \frac{\text{moles of product}}{\text{moles of reactant}} \times \frac{\text{wt. of reactant}}{\text{wt. of one mole of reactant}}$$

36   the mole concept in chemistry

It will be noted that the known chemical equation for the reaction will supply all the necessary information for the right side of this equality, except the weight of reactant specified by the experimental conditions. Hence a simplified form of the generalization is:

$$\text{wt. of product} = \text{wt. of reactant} \times \left[\frac{(\text{no. of moles}) \times (\text{wt. of one mole})_{\text{product}}}{(\text{no. of moles}) \times (\text{wt. of one mole})_{\text{reactant}}} \right]$$

Note that the quantity in the bracket is a dimensionless number (all dimensions cancel). Hence the weight of product will always be expressed in the same units as weight of reactant.

> **EXAMPLE 4.4** What is the gravimetric factor for lime (CaO) in limestone ($CaCO_3$)? Use this to calculate the weight of lime which can be obtained from one ton of crude limestone that assayed at 62.0% of $CaCO_3$.

The equation for the reaction involved is: $CaCO_3 \rightarrow CaO + CO_2$. Hence:

$$\text{wt. of CaO} = \text{wt. of } CaCO_3 \times \left[\frac{1}{1} \times \frac{56.1}{100} \right]$$

$$\underbrace{\text{tons CaO}}_{\text{desired}} = \underbrace{1.00 \text{ ton} \times 0.620}_{\text{given}} \times \underbrace{0.561}_{\substack{\text{gravimetric} \\ \text{factor}}} = 0.348 \text{ tons}$$

The reader can refer to any standard textbook of analytical chemistry to find numerous examples of the utility of this method.[1] We introduce it here to emphasize once again how automatically the working chemist bases his practical methods on the mole concept. Virtually every calculation of chemistry can thus be simplified. However, the beginner is cheating himself if he is willing, at that stage, to accept any method merely because it is "practical" to get an answer quickly. The thinking behind an empirical formula is, for the student, more important than the formula itself.

Calculating Reaction Yields

Most chemical syntheses look better on paper than they do in the laboratory. Many factors are responsible for making the amount of product less than that predicted by calculation from the equation, (the "theoretical yield"). Every article in a chemical journal which reports a new synthesis includes a statement of "% yield" for the benefit of those who may want to use the method described. The yield always is defined relative to the amount of product predicted by the coefficients of the equation thus:

$$\% \text{ yield} = \frac{\text{amount of product obtained}}{\text{amount of product predicted}} \times 100$$

The amount of product is calculated as shown above (Examples 4.1–4.3). Note that the yield, being a dimensionless number can be calculated either on a mole basis or a weight basis.

> **EXAMPLE 4.5** A student performed the synthesis suggested in Example 4.3 and obtained 25.0 g of C_6H_5Br. What was his yield?

---

[1] The classic reference, now in its sixth edition is "Calculations of Analytical Chemistry," by L. F. Hamilton and S. G. Simpson, McGraw Hill Book Co., Inc., New York, 1960. See also the tables provided in such laboratory companions as Lange's "Handbook of Chemistry" or "Handbook of Chemistry and Physics."

$$\% \text{ yield} = \frac{25.0 \text{ (g obtained)}}{50.4 \text{ (g theoretical)}} \times 100 = 49.6\%$$

> **EXAMPLE 4.6** Some of the student's hoped-for product (C_6H_5Br) had reacted with excess Br_2 as follows: $C_6H_5Br + Br_2 \rightarrow C_6H_4Br_2 + HBr$. He recovered 5.0 g of the solid $C_6H_4Br_2$. If this had not occurred, what would his % yield of C_6H_5Br have been? How much benzene (see Example 4.3) remained unreacted?

$$\text{wt. of } C_6H_5Br = 157 \left(\frac{g}{mole}\right)_{C_6H_5Br} \times \frac{1 \text{ (mole)}_{C_6H_5Br}}{1 \text{ (mole)}_{C_6H_5Br_2}} \times \frac{5 \text{ (g)}_{C_6H_5Br_2}}{236 \left(\frac{g}{mole}\right)_{C_6H_5Br_2}}$$

$$y = 3.3 \text{ g of } C_6H_5Br$$

If this 3.3 g had not reacted further, the yield would have been

$$\% \text{ yield} = \frac{25.0 + 3.3 \text{ (g)}}{50.4 \text{ (g theoretical)}} \times 100 = 56.1\%$$

The number of moles of the original benzene (C_6H_6) which did not appear as part of either product is:

$$0.321 \text{ (mole)}_{C_6H_6} \times \frac{100 - 56.1}{100} = 0.141 \text{ mole of } C_6H_6 \text{ unreacted}$$

The following quotation is typical of articles in nearly any issue of the *Journal of the American Chemical Society:*[2] "...0.32 mole of $LiAlH_4$ in ether solution was placed in the flask and 74 g (1.00 mole) of *t*-butyl alcohol was added... The product, $LiAlHC_{12}H_{27}O_3$, weighed 81.0 g."

---

[2] H. C. Brown and R. F. McFarlin, *Jour. Am. Chem. Soc.* **80,** 5375 (1958).

quantitative relationships in chemical reactions 39

EXAMPLE 4.7 If the yield of product is based on the formula given and the amount of $LiAlH_4$ which was reacted, is the author's claim that the yield is "essentially quantitative" a valid one?

It can be seen by inspection of the formulas that one mole of $LiAlH_4$ should theoretically ("quantitatively") produce one mole of $LiAlHC_{12}H_{27}O_3$. If this is so, then 0.32 mole should produce 0.32 mole of product. The formula weight of the product is 254. Hence $\dfrac{81.0 \text{ (g)}}{254 \left(\dfrac{g}{mole}\right)} = 0.32$ mole is 100% yield.

EXAMPLE 4.8 A student took 15 g of a fat and converted it into soap by the equation:

$$\begin{array}{c}
\text{H} \\
\text{HC—OCOR} \\
| \\
\text{HC—OCOR} \\
| \\
\text{HC—OCOR} \\
\text{H}
\end{array} + 3\text{NaOH} \rightarrow \begin{array}{c}
\text{H} \\
\text{HC—OH} \\
| \\
\text{HC—OH} \\
| \\
\text{HC—OH} \\
\text{H}
\end{array} + 3\text{NaOCOR}$$

fat   *soap*

The formula for soap, written with "R" in this way, implies that uncertain or *average* composition prevails. Nevertheless, it is possible by other experiments to evaluate an average formula weight for the fat. This was found to be 830. If a student should get a yield of 75%, how much soap can he expect?

The first problem is to calculate a value for the formula weight of "R." If we write the formula for the fat as $C_6H_5O_6R_3 = 72 + 5 + 96 + 3R = 830$, we can solve for

40 the mole concept in chemistry

R = (830 − 173)/3 = 219. Using this value, we find that the formula weight for soap is 286. Then

$$y \text{ g of soap} = 286 \left(\frac{g}{\text{mole}}\right)_{\text{soap}} \times \frac{3 \text{ (mole)}_{\text{soap}}}{1 \text{ (mole)}_{\text{fat}}} \times \frac{15 \text{ (g)}_{\text{fat}}}{830 \left(\frac{g}{\text{mole}}\right)_{\text{fat}}}$$

$$y = 15.5 \text{ g soap for } 100\% \text{ yield}$$
$$10.6 \text{ g for } 75\% \text{ yield}$$

Calculating Amounts of Gases Reacting

It is clear from the development outlined in Chapter 2 that whenever gases are involved, the chemist can measure the number of moles by other means than by weighing. When a gaseous species is reacting according to a known equation, the first consideration is exactly the same as if it were a solid or a liquid: how many moles of it are involved? Often problems are posed in terms of a volume of a gas (known P and T). This is only a special case; the problem could just as well be set up to ask for a pressure of a gas (known V and T). The important relationship is that stated at the beginning of this discussion:

$$\begin{array}{c} \text{number of moles} \\ \text{of a gas product:} \\ \text{calculated by } PVT \end{array} = \frac{\begin{array}{c}\text{ratio from} \\ \text{equation} \\ \text{moles of product}\end{array}}{\text{moles of reactant}} \times \begin{array}{c}\text{number of moles} \\ \text{of reactants}\end{array}$$

EXAMPLE 4.9 Reconsider Example 4.1. What volume of oxygen (STP) will be required to react with 1.0 g of calcium by the equation:

$$\text{Ca} + \tfrac{1}{2}\text{O}_2 \rightarrow \text{CaO}$$

$$\text{moles of O}_2 = \frac{0.5 \text{ (mole)}_{\text{O}_2}}{1.0 \text{ (mole)}_{\text{Ca}}} \times \frac{1.0 \text{ (g)}_{\text{Ca}}}{40.0 \left(\frac{g}{\text{mole}}\right)_{\text{Ca}}} = 0.0125 \text{ mole}$$

quantitative relationships in chemical reactions

If the answer is to be expressed as a volume, the ideal gas equation of state is used to translate the fundamental quantity, number of moles, into the desired volume;

$$V = n \frac{RT}{P}$$

$$\text{Vol. of } O_2 = \frac{0.0125 \text{ (mole)} \times 0.0820 \left(\frac{1 \text{ atm}}{\text{mole} \,^\circ K}\right) \times 273 \,(^\circ K)}{1.00 \text{ (atm)}} = 0.280 \text{ l}$$

This special case (calculating STP volume) is made even simpler by using 22.4 l, the STP volume of one mole. Thus:

$$\text{Vol. of } O_2 = 0.0125 \text{ (mole)} \times 22.4 \left(\frac{l}{\text{mole}}\right) = 0.280 \text{ l (STP)}$$

The more general case is illustrated by data involving other than STP conditions. Here again, we emphasize the significance of using the ideal gas law directly, either to calculate the number of moles or using the number of moles to calculate the predicted value of some other variable. Consider the example:

EXAMPLE 4.10 A student knows that hydrogen gas can be generated from an acid by reaction with magnesium:

$$Mg + 2H_3O^+ \rightarrow H_2 + Mg^{++} + 2H_2O$$

What is the maximum weight of magnesium he can use if he expects to collect the H_2 gas in a 50 ml buret on a day when the barometer is 740 mm Hg? He probably will collect the gas over water at 15° C, so that it will contain about 13 mm Hg of water vapor.

The equation coefficients for Mg and H_2 are 1 to 1 so

| x mole of Mg will produce (calculated by weight) | $\longleftrightarrow$ | x mole of H_2 (calculated by PVT) |

$$\frac{y \text{ g of Mg}}{24.3 \left(\frac{\text{g}}{\text{mole}}\right)} = x \text{ mole} = \frac{\frac{740-13}{760} \text{ (atm)} \times 0.050 \text{ (l)}}{0.0820 \left(\frac{1 \text{ atm}}{\text{mole} \, ^\circ\text{K}}\right) \times 288 \, (^\circ\text{K})}$$

$$y = \frac{0.957 \times 0.050}{0.082 \times 288} \times 24.3 = 0.0492 \text{ g of Mg}$$

It is possible to determine the relative amounts of two substances in a mixture if they are known to react to form a common product. An example is the following:

> **EXAMPLE 4.11** A student treated a sample of an alloy of zinc and aluminum weighing 0.156 grams with acid and collected the hydrogen gas evolved. 144 ml of H_2 gas were collected over water at 20° C. (vapor pressure = 17 mm Hg) at a barometric pressure of 742 mm Hg. What is the composition of the alloy?[3]

If x g = weight of Zn in the sample, $(0.156 - x)$ g = wt. of Al.

$$\text{Then moles of Zn in sample} = \frac{x \text{ (g)}}{65.4 \left(\frac{\text{g}}{\text{mole}}\right)}$$

$$\text{moles of Al in sample} = \frac{(0.156 - x) \text{ (g)}}{27.0 \left(\frac{\text{g}}{\text{mole}}\right)}$$

total moles of H_2 produced are:

$$n = \frac{\frac{742-17}{760} \text{ (atm)} \times 0.144 \text{ (l)}}{0.0820 \left(\frac{1 \text{ atm}}{\text{mole} \, ^\circ\text{K}}\right) \times 293 \, (^\circ\text{K})} = 0.00572 \text{ mole}$$

[3] See W. T. Masterton, *Jour. Chem. Educ.* **38,** 558 (1961).

The equations for reaction are:

$$Zn + 2H_3O^+ \rightarrow H_2 + Zn^{++} + 2H_2O$$

$$Al + 3H_3O^+ \rightarrow \tfrac{3}{2}H_2 + Al^{+++} + 3H_2O$$

This means that

$\dfrac{x \text{ (g)}}{65.4 \left(\frac{g}{\text{mole}}\right)}$ mole of Zn will produce $\dfrac{x}{65.4}$ mole of H_2 and

$\dfrac{(0.156 - x) \text{ (g)}}{27.0 \left(\frac{g}{\text{mole}}\right)}$ mole of Al will produce $\dfrac{3}{2} \times \dfrac{0.156 - x}{27}$ mole of H_2

The total number of moles of H_2 has been established from PVT data so:

total moles of H_2	=	moles of H_2 from Zn reaction	+	moles of H_2 from Al reaction
0.00572	=	$\dfrac{x}{65.4}$	+	$\dfrac{1.5(0.156 - x)}{27}$
0.00572	=	$0.0153x$	+	$0.00866 - 0.0556 x$
x	=	0.073		

The alloy is $\dfrac{0.073 \text{ (g)}_{Zn}}{0.156 \text{ (g)}_{Zn + Al}} \times 100 = 47\%$ Zn

This same interpretation can hold for any similar circumstance in which it is possible to count by any means the total moles of reaction produced by a binary (two component) mixture and at the same time to represent the number of moles of each component by only one unknown.

If a gas is a product in a chemical reaction, a chemist almost always measures it by PVT data, but in a report of his research, he probably takes for granted that his readers realize this. For the sake of economy in publication space, he will report his work in terms of moles (or for small amounts, the unit "*milli-*

moles" is preferred). The following example is typical, reported in the same article from which Example 4.7 was chosen:[4]

> **EXAMPLE 4.12** "52.5 millimole of $LiAlH_4$ was treated with 15.6 g (210 millimole) t butyl alcohol... A total of 156 millimole of hydrogen was evolved for the reaction:
>
> $LiAlH_4 + 3(CH_3)_3COH \rightarrow 3H_2 + Li[(CH_3)_3O]_3AlH\ldots$"
>
> The addition of an excess of another alcohol, methanol, to the above reaction mixture caused the fourth H atom of the $LiAlH_4$ to be replaced according to the equation:
>
> $Li[(CH_3)_3O]_3AlH + CH_3OH \rightarrow H_2 + Li[(CH_3)_3O]_3[CH_3O]Al$
>
> How much H_2 was evolved?

Note the economy of statement to convey a maximum of information. The ratio: 156 millimole/52.5 millimole tells the reader to expect the first reaction to be represented by an equation with coefficients 3 for H_2 to 1 for $LiAlH_4$. Furthermore, even without understanding the details of how the reaction occurred or even without knowing the formula of t butyl alcohol the reader knows that an excess of it was added. (210 − 3 × 52.5 = 52.5 millimole excess).

When the second alcohol is added, the expected amount of hydrogen will be in a ratio of 1 mole of H_2 to 1 mole of the original $LiAlH_4$. Thus, the answer expected is 52.5 millimole since this is the amount of starting material, $LiAlH_4$. (The reader can refer to the published article to see that the experimental value, 51 millimole, was close to his prediction.) We hope that the reader recognizes the point of this example: what may appear to be complicated chemistry is complicated only in detail, not in essential representation.

When all species in a chemical reaction are gases, the prob-

[4] See Footnote 2, p. 38.

lem of measuring relative numbers of moles is very simple. If conditions of constant pressure and temperature can be maintained, the situation is that described by Gay-Lussac's Law of combining gas volumes. For example, if a chemist has one cubic whiffledink of oxygen gas which he hopes to convert completely to ozone by the equation: $3O_2 \rightarrow 2O_3$, he knows that he can expect 2/3 of a cubic whiffledink of product. Although he has no idea (or way of discovering) how many moles are in one cubic whiffledink, if P and T are unchanged, the number will be constant. Hence, the direct proportionality between numbers of moles and gas volumes allows him to translate equation coefficients directly into volumes. In actual practice, problems of gas analysis are based on this simple principle. Often it is more convenient to measure changes in the pressure of a constant-volume sample before and after reaction.

Gas phase reactions produced by absorption of high energy radiation are often studied on a micro scale by adapting the simple principle emphasized above. One such technique is described by Saunders and Taylor.[5] It consists of measuring the changes in pressure as a consequence of freezing out constituents in the mixture, or combining $H_2 + O_2$ to form H_2O on a hot platinum filament, etc. The following are typical data:

EXAMPLE 4.13 5.22×10^{-4} mole of a gas known to contain H_2, O_2, and N_2 exerted a pressure of 67.4 mm Hg in a certain standard volume. The gas was passed over a hot platinum filament which combined H_2 and O_2 into H_2O which was frozen out. When the gas was returned to the same volume, the pressure was 14.3 mm. Extra oxygen was added to increase the pressure to 44.3 mm. The combustion was repeated, after which the pressure read 32.9 mm. What was the composition of the gas sample?

[5] K. W. Saunders and H. A. Taylor, *J. Chem. Phys.* **9**, 616 (1941).

46 the mole concept in chemistry

This is another example to demonstrate how apparently complicated information can fit into a very simple interpretation. The combusion reaction is:

$2H_2 + O_2 \rightarrow$ a product which exerts no gas pressure
3 mole of gas $\rightarrow$ 0 mole of gas.

Since V and T are constant, $n = P \times \dfrac{V}{RT} = P \times$ constant. So, changes in pressure can be interpreted as numbers of moles. Note that the amount of H_2 in the mixture must have exceeded twice the amount of O_2 originally present, since after the first combustion, a second with added O_2 produced a pressure drop. We leave it to the reader to supply the reasoning behind the following:

$$\text{mole fraction of } H_2 = \frac{\frac{2}{3}(67.4 - 14.3) + \frac{2}{3}(44.3 - 32.9)}{67.4} = 0.638$$

$$\text{mole fraction of } O_2 = \frac{\frac{1}{3}(67.4 - 14.3)}{67.4} = 0.262$$

$$\text{mole fraction of } N_2 = 0.100$$

chapter five

PROPERTIES OF LIQUID SOLUTIONS

One definition of a solution frequently found in textbooks is the semantic contradiction, a "homogeneous mixture." The contradiction loses its force when we recognize that we are thinking "homogeneous" at one level (the macroscopic) and "mixture" at another (the microscopic). The solution does behave homogeneously when tested for uniformity of physical and chemical properties. The uniformity revealed by these measurements is the uniformity of mixture composition at the particulate level of molecules, ions, etc. Hence it is natural to expect that the mole idea will be a useful concept to interpret relationships between the magnitude of values for macroscopic properties and the composition of solutions. We shall limit our discussion to solutions in which the resulting phase is a liquid. Solutions of gases in gases are nothing more than mixtures of gases; these we have already treated in Chapter 2. Solid solutions are less susceptible to simple treatment, though to a first approximation they often can be treated similarly to liquid solutions.

Before the properties of liquid solutions are discussed, it may be well to organize our thinking about the properties of liquids in general. Most of the important information about liquid solutions can be derived from properties measured in comparison with those of the pure constituents, usually of the solvent. The dependence of these properties on relative numbers of solvent and solute particles (molecules, ions, etc.) will be our main concern. There are two kinds of properties of any substance, extensive and intensive. Let us examine each category.

Extensive properties are those which depend not only on what the substance is, but on how much of it is being considered. For liquids, typical important extensive properties are weight (mass), volume, heat capacity, etc. Since these depend on the amount of material in the sample, any measurement must be calculated in terms of a "specific" value, that is the value of a measurement made on a definite amount, usually one gram. Thus the specific volume of liquid water is 1.00177 ml g^{-1} at 20°C; its specific heat is 1.00 cal g^{-1} at 15°C, etc. The chemist usually finds his calculations simplified by using values for the measurements made on one mole of the material. The reasoning behind this is exactly the same as that inherent in his choosing the mole in the first place—namely, his knowing the value for the same number of chemical units of comparable substances. Thus he refers to the "molar" volume of water as 18.00319 ml mole^{-1}, the molar heat capacity as 18.00 cal mole^{-1} (°K)$^{-1}$. Whenever any inferences involving the character of the species are to be drawn, it is necessary to start with values that measure a property of the same number of species. The following example is suggestive of hundreds of comparisons that can be made, comparisons which can serve as the basis for ultimate calculation of molecular properties.

EXAMPLE 5.1 The heat of vaporization of H_2O is 540 cal g^{-1} for heavy water (D_2O) 497 cal g^{-1}, both at the normal boiling points. Which substance requires more heat to change one mole from liquid to vapor? *Answer:* D_2O: 9940 cal, compared to 9720 cal for H_2O.

In general, the extensive properties of liquid solutions reveal less information conveniently than do the intensive properties. If there is no interaction between the different particles of a solution, it is to be expected that the additive nature of extensive properties will be reflected in the properties of the mixtures (the solutions). Thus the weight of a solution is the sum of the weights of the constituents. Volumes are less likely to be additive, because the bulk volume of a liquid often contains holes or vacant spaces resulting from the randomness and irregularity of aggregation of the molecules in any given liquid. Consequently, a liquid may accomodate some of the "invader" particles of another substance by having the latter "tucked into" existing vacancies and hence not reflect their presence by a total volume change.

A second kind of property is the intensive type. These properties depend not at all on how much substance is present. Melting point, boiling point, vapor pressure, refractive index, etc., are examples. These are important physical constants for any particular species and hence their observation can serve for qualitative identification. There is no such thing as a "molar boiling point" because the temperature at which a liquid boils is completely independent of the amount of material (number of moles) under scrutiny.[1]

However, when two or more pure substances are mixed into a solution, we find that the intensive properties of any one substance are influenced by the presence of the others. Just as it is possible to identify a pure substance by a knowledge of its intensive properties, it becomes possible to evaluate its concentra-

---

[1] We must be careful not to push this statement too far. It is not correct, for example, to speak of the boiling point of one molecule of a substance. We must limit our consideration to a sample size such that substantially all of the molecular units present are in the same environment with respect to numbers of neighbors. Even on a macro scale some variations are measurable; for example, the differences in vapor pressure from a curved liquid surface inside a bubble. See for example: E. F. Hammel, *Jour. Chem. Educ.* **35**, 28 (1958).

tion in a mixture by measuring the change in those intensive properties. For example, the vapor pressure of a liquid is a manifestation of the escaping tendency of its molecules from the liquid to the gas phase at a given temperature. If now, these molecules in the liquid phase are uniformly mixed with others, it is to be expected that the number of any one kind at the liquid-gas interface will be relatively less than if the substance were pure. Consequently, a new factor, relative number of molecules, has been introduced to influence the escaping tendency.

We have chosen escaping tendency, more practically called vapor pressure, as our example deliberately. Raoult, in 1886, enunciated the law which is the cornerstone for interpretations of the properties of solutions: $P_A = P_A^\circ \times X_A$. This states that the vapor pressure of a liquid A over a solution in which it is present will be proportional to its concentration expressed as the mole fraction X_A. The constant of proportionality, P_A°, is the vapor pressure of the pure liquid at that temperature. Since $X_A < 1.0$, P_A will always be less than P_A°. Raoult's Law implies, then, that the *only* factor influencing the escaping tendency of the molecules of a given species is their relative number. If these conditions are true, a solution is said to be ideal. Actually, there are few cases in which the molecules of liquid A are not influenced by B molecules or any other admixed type of particles. However, we can readily see that if a solution is very dilute, the molecules of the solvent, being in much greater abundance will be little influenced by the few solute invaders. Consequently, Raoult's Law predicts quite well the vapor pressure of a solvent when its mole fraction is large; i.e., when the solution is dilute.

We shall limit our discussion to applicability of Raoult's Law to the solvent in dilute solutions containing non-volatile solutes. Sugar in water is such a solution.

If Raoult's Law is written:

properties of liquid solutions

$$P°_{solvent} - P_{solvent} = P°_{solvent} \times (1 - X_{solvent})$$

the mole fraction of solute will be $(1 - X_{solvent})$ so:

$$\Delta P_{solvent} = P°_{solvent} \times X_{solute}$$

This important relation suggests that the decrease in solvent vapor pressure can be used to measure the amount of solute.

The following example is chosen to make several points clear. First is the definition and calculation of mole fractions. Second is the application of Raoult's Law. Third is the more subtle but equally important choice of the *form* of the law to obtain an answer of maximum significance with a minimum of arithmetical manipulation.

EXAMPLE 5.2 The vapor pressure of H_2O is 23.756 mm Hg at 25.0° C. What is the vapor pressure of a solution of 28.5 g of sucrose ($C_{12}H_{22}O_{11}$) in 100 g of H_2O?

The numbers of moles of each species:

$$n_{H_2O} = \frac{100 \text{ (g)}}{18.0 \left(\frac{g}{mole}\right)} = 5.55;$$

$$n_{C_{12}H_{22}O_{11}} = \frac{28.5 \text{ (g)}}{342 \left(\frac{g}{mole}\right)} = 0.0834$$

The mole fraction of each species:

$$X_{H_2O} = \frac{5.55}{5.55 + 0.08} = 0.985$$

$$X_{C_{12}H_{22}O_{11}} = \frac{0.0834}{5.55 + 0.08} = 0.0148$$

If Raoult's Law is used in the form:

$$P_{H_2O} = P°_{H_2O} \times X_{H_2O}$$

$$P_{H_2O} = 23.756 \times 0.985 = 23.4(0) \text{ mm Hg}$$

In the alternate form:

$$\Delta P_{H_2O} = P°_{H_2O} \times X_{C_{12}H_{22}O_{11}}$$

$$\Delta P_{H_2O} = 23.756 \times 0.0148 = 0.352 \text{ mm Hg}$$

$$P_{H_2O} = 23.756 - 0.352 = 23.404 \text{ mm Hg}$$

It will be noted that the latter method gives an answer of five significant figures, whereas the former is good to only three. This should sharpen the reader's appreciation for the sensitivity of any method which calculates differences between large numbers.

Colligative Properties

There are various other intensive properties of liquids directly related to the vapor pressure. One of these, the normal boiling point, is defined as the temperature at which the vapor pressure will equal an opposing pressure of one atmosphere. The freezing point also has a connection, though it is less obvious from the usual definition. We can say, however, that the freezing point is that temperature at which the vapor pressure of a liquid equals that of the pure solid phase. That this is true can be seen if one imagines an inequality of escaping tendencies from solid and liquid when samples are put in a closed container with no other molecules present. Solid would distill to liquid if the escaping tendency of the molecules of the solid was the greater of the two. Equilibrium can be attained only if the molecules tend to leave and return to all phases equally.

The consequence of lowering the solvent vapor pressure by the presence of solute particles is that the solution will boil at a higher temperature and freeze at a lower temperature than would be characteristic of the pure solvent substance. Raoult's Law shows that the vapor pressure decrease, $\Delta P_{solvent}$, is proportional to the number of solute particles, X_{solute}. We can then expect that boiling point rise, ΔT_b, or freezing point depression, ΔT_f, will also be proportional to X_{solute}. These are two of the colligative properties of solutions.

It is a simple matter to measure a temperature very accurately, especially when what is wanted is only a difference (ΔT) from an easily established calibration point such as the freezing or boiling temperature of a pure solvent. Consequently, the chemist turns to either of these measurements, boiling-point rise or freezing-point lowering to tell him about the concentration of a solution. The only standard of convenience that remains is to establish conditions such that ΔT will be proportional to some easy, direct measure of number of solute particles.

These considerations have led to the establishment of the *molal* concentration scale. The *molality* of a solution is defined as the number of moles of solute per 1000 g of solvent. We have

$$\Delta T_b = K_b m$$

$$\Delta T_f = K_f m$$

where K_b and K_f are constants characteristic of the pure solvent substance and ΔT_b and ΔT_f are the changes in the boiling point and freezing point, respectively, produced by introducing a solute of molality m.

The astute reader will realize that we have moved a little too fast under the guise of "convenience." Raoult's Law stated that ΔP was proportional to X_{solute}. Even if thermodynamics can show that ΔP is directly proportional to ΔT_b or ΔT_f, we certainly cannot rewrite Raoult's Law: $\Delta P \propto X_{solute} \propto \Delta T \propto$ molality, because molality and mole fraction are defined differently. However, if we can show mathematically that changes in the molality are very nearly in direct proportion to changes in the mole fraction, we can expect the equation, $\Delta T_{b\,or\,f} = K_{b\,or\,f}m$, to hold. The following numerical example demonstrates this very real limitation on the applicability, no matter how "convenient" the equation.

> EXAMPLE 5.3 Calculate and compare the mole fractions of solutes in 0.25, 0.50 and 1.00 molal aqueous solutions. Repeat the calculations for the same molalities in the solvent CCl_4.

A molality of 0.25 means that 0.25 mole of solute is present in 1000 g, $1000/18.0 = 55.5$ mole of H_2O. Correspondingly in CCl_4, we have 0.25 mole of solute in $1000/153.8 = 6.50$ mole. See tabulation below.

Molality	X_{solute} in H_2O	X_{solute} in CCl_4
0.25	$\dfrac{0.25}{55.55 + 0.25} = 0.00448$	$\dfrac{0.25}{6.50 + 0.25} = 0.0370$
0.50	$\dfrac{0.50}{56.05} = 0.00892$	$\dfrac{0.50}{7.00} = 0.0714$
1.00	$\dfrac{1.00}{56.55} = 0.0177$	$\dfrac{1.00}{7.50} = 0.133$

Comparison of the numbers is very revealing. In aqueous solutions we find that the ratio of the molalities is very nearly equal to the ratio of the solute mole fractions. (e.g. $0.25/1.00 = 0.250$; $0.00448/0.0177 = 0.253$.) The same is not true for the CCl_4 solutions ($0.0370/0.133 = 0.278$). It should be apparent that this is not the consequence of any lack of ideality on the part of CCl_4 solutions. Rather it is the important mathematical consequence of the definition of molality

properties of liquid solutions 55

based upon a definite *weight* of solvent (1000 g) rather than on a definite number of moles of solvent. A more general conclusion is that we can expect the equation $\Delta T_{b\,or\,f} = K_{b\,or\,f} m$ to hold only when the solute mole fraction is of the order of 0.01 or less. This limitation further underscores the limitation imposed by considering that Raoult's Law (upon which the whole derivation is based) is valid only for dilute solutions in the first place!

If we agree, then, not to expect too much from the equation $\Delta T_{b\,or\,f} = K_{b\,or\,f} m$, we can see how useful it is. Here is one more technique for counting the number of moles in a sample of a pure material: put it into a solvent for which $K_{b\,or\,f}$ is known and measure the corresponding $\Delta T_{b\,or\,f}$. Accordingly, if this reveals the number of moles in a known weight, the weight of one mole can be calculated. The more common technique is to measure the depression of a solvent freezing point. This is referred to as cryoscopy (from the Greek word for cold).

EXAMPLE 5.4 K_f for water is 1.86 (°C molality^{-1}). What is the molality of a solution which freezes at -0.192°C? Assuming no change in the solute by raising the temperature, at what temperature will the solution boil? (K_b for H$_2$O = 0.515 (°C molality^{-1}))

$$\Delta T_f = K_f m$$

$$\text{molality} = \frac{-0.192}{-1.86} = 0.103 \text{ molal}$$

$$\Delta T_b = 0.515 \times 0.103 = 0.0532 \text{ (°C)}$$

If the barometer = 760 mm Hg, $T_b = 100.053°$ C

EXAMPLE 5.5   What is the weight of one mole of a solute, 0.132 g of which in 29.70 g of benzene gave a freezing-point depression of 0.108° C? K_f for benzene is 5.12.

This problem is essentially similar to that in which the weight of one mole of a gas was calculated. In both the approach is to calculate a number of moles by two different methods (Chapter 2, Example 2.4). The resulting equality is the equation which can be solved for the unknown quantity. In this case we calculate the number of moles of solute in 1000 g of solvent (the molality):

$$\frac{\Delta T_f}{K_f} = m = \frac{g}{M} \times \frac{1000}{G}$$

$$\frac{\text{Observed } \Delta T}{\Delta T \text{ for 1 mole of solute in 1000 g of solvent}} = \begin{array}{c}\text{number of moles}\\\text{of solute in}\\ G \text{ g of solvent}\end{array} \times \begin{array}{c}\text{factor converting}\\\text{to basis of}\\ 1000 \text{ g of solvent}\end{array}$$

Here g is the measured solute weight, G the measured solvent weight, and M (g mole^{-1}) the weight of one mole of solute. The unknown in the example is M. So:

$$\frac{0.108 \, (°C)}{5.12 \, (°C \text{ kg mole}^{-1})} = \text{molality} = \frac{0.132 \, (g)}{M \, (g \text{ mole}^{-1})} \times \frac{1000 \, (g \text{ kg}^{-1})}{29.70 \, (g)}$$

$$M = \frac{0.132 \times 1000 \times 5.12}{0.108 \times 29.70} = 211 \text{ g mole}^{-1}$$

The reader can easily find examples in which some other of the quantities in this relationship appears as the unknown. Every textbook of general chemistry contains them.

The organic chemist finds cryoscopy to be one of his most useful tools in knowing what to assign as a formula weight for an unknown material. His unknowns usually are compounds insoluble in water. Further, he often needs to know an approximate value quickly and without tying up elaborate equipment. A method suggested first by Rast meets these criteria admirably by using camphor (melting point 179°C) as a solvent. The abnormally high K_f, 39.7 (°C molality^{-1}), means that even though small solute samples are used, ΔT will be so large that measurement with an ordinary thermometer will give sufficient precision for approximate work. However, K_f is by no means constant at low molalities of solute (below the range of 0.2–0.5 molal).[2] Hence, it is often necessary to run an analysis twice, the first time to get an approximate value, the second time to choose a sample on the basis of an "educated guess" such that the solute concentration will be in the 0.2–0.5 molal range where the accepted 39.7° value for K_f can be used.

> EXAMPLE 5.6 A student took 0.100 g of an unknown, dissolved it in 5.00 g of camphor, and found that the camphor melting point was depressed 5.3° C. If he uses the value 39.7 for K_f, what is the approximate weight of one mole of solute? Then: what size sample should he choose to be sure that using $K_f = 39.7$ will give him a correct weight per mole? What is the correct weight of one mole of his unknown?

The approximate weight of one mole can be found by:

$$\frac{\Delta T_f}{K_f} = \text{molality} = \frac{g}{M} \times \frac{1000}{G}$$

$$\frac{5.3 \, (°C)}{39.7 \left(\frac{°C}{\text{molality}}\right)} = \frac{0.100 \, (g)}{M \left(\frac{g}{\text{mole}}\right)} \times \frac{1000 \, (g)}{5.00 \, (g)}$$

$$M \approx 150 \text{ g mole}^{-1}$$

[2] See W. B. Meldrum, L. P. Saxer, and T. O. Jones, *Jour. Am. Chem. Soc.* 55, 2023 (1943).

58 the mole concept in chemistry

An approximately 0.2 molal solution will require in 5.00 g of solvent:

$$m = 0.200 = \frac{g}{150} \times \frac{1000}{5.00}$$

$$g = 0.200 \text{ (g)} \times 150 \left(\frac{g}{\text{mole}}\right) \times \frac{5.00 \text{ (g)}}{1000 \text{ (g)}}$$

$$g = 0.150 \text{ g}$$

When a sample of this weight was used, the observed freezing point depression was 7.73°.

$$\frac{7.73 \text{ (°C)}}{39.7 \left(\frac{°C}{\text{molality}}\right)} = \frac{0.150 \text{ (g)}}{M \left(\frac{g}{\text{mole}}\right)} \times \frac{1000 \text{ (g)}}{5.00 \text{ (g)}}$$

$$M = 154 \text{ g mole}^{-1}$$

Non-ideal Behavior of Solutes

The usefulness of observing colligative properties of solutions goes far beyond merely evaluating the weight of one mole of a purified solute. The properties of the solvent change in response to the presence of all solute particles, regardless of their type. If, for example, an aqueous solution contains 0.10 mole of sugar, plus 0.15 mole of urea in 1000 g of H_2O, the total molality of solute will be:

$$m = (0.10 + 0.15) = \frac{\Delta T_f}{K_f}$$

and the freezing point will be $0.25 \times 1.86° = 0.465°$ lower than the freezing point of pure water.

The cryoscopic method frequently is used effectively to count the moles of solute particles and compare the result with the number of moles of solute anticipated by use of the formula weight. A typical example of the application of this line of reasoning was its use to help elucidate the structure of inor-

properties of liquid solutions

ganic complex compounds. Certain transition metal salts, such as cobalt chloride, are found to crystallize into a variety of forms when precipitated from aqueous solutions of ammonia. The monumental researches of Alfred Werner at the start of the present century provided a consistent picture of complexation by ascribing two "spheres of attraction" to a central ion. The inner sphere, termed the coordination sphere, held the adducts tightly; the second sphere allowed the units to break away readily. For example, the compound which can be isolated and shown to have the empirical formula $CoCl_3 \cdot 6NH_3$ has nine units ($3Cl^- + 6NH_3$) "fastened" somehow to the central Co^{+3}. Werner postulated the structure to be $Co(NH_3)_6^{+++}$, $3Cl^-$. Such evidence as the following gives him strong support.

EXAMPLE 5.7 1.1 g of $CoCl_3 \cdot 6NH_3$ (formula weight = 267) were dissolved in 100 g of H_2O. The freezing point of the solution was $-0.29°C$. How many moles of solute particles exist in solution for each mole of solute introduced?

If we calculate the number of moles of solute particles per 1000 g of H_2O solvent from the cryoscopic data, m_C, we have:

$$m_C = \frac{\Delta T_f (°C)}{K_f \left(\frac{°C}{\text{molality}}\right)} = \frac{0.29}{1.86} = 0.156 \text{ mole of solute particles}$$

The same quantity, m_F, calculated on the basis of the formula is:

$$m_F = \frac{g}{M} \times \frac{1000}{G} = \frac{1.1 \text{ (g)}}{267 \left(\frac{g}{\text{mole}}\right)} \times \frac{1000 \text{ (g)}}{100 \text{ (g)}} = 0.0412 \text{ molal}$$

The ratio:

$$\frac{m_C}{m_F} = \frac{0.156}{0.0412} \approx 4$$

indicates that the solvent behaves as if each formula mole of complex salt exists in solution as four solute particles. Other data support this view (see Chapter 7, page 88).

Observations on the freezing points of aqueous solutions of salts are of tremendous historical significance. This was one line of evidence which led the physical chemists of the late 19th century to postulate that "molecules" of salts dissociated into electrically charged ions. For example, if 5.35 g of NH_4Cl (0.10 mole calculated on the basis of the formula weight = 53.5) was dissolved in 1000 g of water, the freezing point was observed to be 0.344° C below that for pure water. Now we notice that our approach of calculating the "effective" number of moles of solute:

$$\frac{\Delta T_f}{K_f} = m = \frac{0.344}{1.86} = 0.185 \text{ molal}$$

does not agree with the figure based upon the weight of one mole established by the formula:

$$\frac{5.35 \text{ (g)}}{53.5 \left(\frac{g}{mole}\right)} = 0.100 \text{ molal}$$

Arrhenius proposed that the following had happened:

	(NH_4Cl)	$\rightleftharpoons$	NH_4^+	+	Cl^-
Amounts put into solution	0.100 mole	+	0.0 mole	+	0.0 mole
Amounts present at equilibrium	$0.100(1 - \alpha)$ mole	+	0.100α mole	+	0.100α mole

Total moles of solute particles = $0.100 (1 + \alpha)$

Here α appears as a degree of dissociation. If we now equate the number of moles of solute:

$$\underbrace{\frac{\Delta T_f}{K_f} = \frac{0.344}{1.86}}_{\text{established experimentally}} = \underbrace{0.100 (1 + \alpha)}_{\text{represented by dissociation equation}}$$

We can calculate α, the degree of dissociation:

$$\alpha = \frac{0.185 - 0.100}{0.100} = 0.85$$

The modern theory of electrolytes recognizes no such concept as a "molecule" of a salt. Rather it postulates that ions are the only species in aqueous salt solution. This would predict, in the NH_4Cl case just discussed, that the freezing-point lowering should be

$$\Delta T_f = 2 \times m \times K_f = 0.372°\,C$$

instead of the observed $\Delta T_f = 0.344°\,C$. The solution has not responded as though it knew the theory. The solvent still acts as though it "sees" only 0.185 mole of particles of *any* kind. The way out of the apparent contradiction is to recognize that interactions between solute particles, especially charged ions, always will in turn affect the behavior of the solvent particles. This leads to the concept of "activity" (a) as a substitute for "concentration" (c) of species and the empirical use of an "activity coefficient" (f) defined as a/c. Current research in this field is directed to the correlation of colligative-property data by means of this and other related concepts.

Molecular associations frequently can be revealed by colligative property studies. The following is a typical example. However, the astute reader will recall the discussion immediately following Example 5.3. He will then recognize that the answer can be only an approximation.

EXAMPLE 5.8 J. H. Simons and M. G. Powell (*Jour. Am. Chem. Soc.* **67**, 77 (1945)) found that the freezing point of CCl_4 was depressed by $5.415°\,C$ when 60.26 g of VCl_4 was added to 1000 g of solvent CCl_4. Simons and Powell used $K_f = 29.9$. Compare the number of moles of particles with the number predicted by the formula. Calculate the number of dimers, V_2Cl_8, present.

"effective" molality $= \dfrac{\Delta T_f (°C)}{K_f \left(\dfrac{°C}{\text{molality}}\right)} = \dfrac{5.415}{29.9} = 0.181$ molal

"formula" molality $= \dfrac{\text{weight (g)}}{M \left(\dfrac{g}{\text{mole}}\right)} = \dfrac{60.26}{192.7} = 0.315$ molal

The fact that the solvent "sees" fewer particles than anticipated from the formula means that association has occurred. (This might be expected upon realizing that a vanadium atom probably has an odd electron left after bonding in VCl_4.) If we postulate:

$$2VCl_4 \rightleftharpoons V_2Cl_8$$

Amounts put into solution	0.315 mole	+	0.0 mole
Amounts present at equilibrium	$0.315(1 - 2\beta)$	+	0.315β

total moles of solute particles $= 0.315(1 - \beta)$

Here 2β is the fraction of the VCl_4 molecules which have dimerized. The total moles of particles measures the "effective molality."

Hence:

$$\underbrace{0.181}_{\substack{\text{established} \\ \text{experimentally}}} = \underbrace{0.315(1 - \beta)}_{\substack{\text{represented by} \\ \text{dimerization equation}}}$$

$$\beta = 0.425$$

The moles of each species present will be: $V_2Cl_8 = 0.315 \times 0.425 = 0.134$ mole; $VCl_4 = 0.315(1 - 2 \times 0.425) = 0.0473$ mole.

chapter six

CHEMICAL EQUIVALENCE

All chemical knowledge rests on the chemist's ability to measure the amounts of substances. This is the business of analytical chemistry. Fundamentally, what is done is to apply the principle set forth in Chapter 4, "Quantitative Relationships from Chemical Equations." The amount of a species is revealed by the known amount of something with which it reacts. Sometimes, especially when the mechanism of a complex reaction is being investigated, the species present can only be postulated since their anticipated lifetime may be only transitory. At other times the amounts of any one species may depend on the simultaneous presence of various others, all in a state of dynamic chemical equilibrium. Nevertheless, ultimately a knowledge of what is going on in a chemical reaction must depend on our being able to measure the number of moles of a chemical species on the basis of an assigned formula. The beginning student should not lose sight of this simple fact when he reads about the mysterious complexity of modern methods of analysis: spectroscopy, polarography, chromatography, etc. These are merely elaborations of the essential argument we shall present in this chapter.

The foundation for any analytical method is the standard sample. Some methods allow the use of known weights of pure materials. It is obvious from our previous discussion, that what the chemist invariably does is to translate a given weight into a number of moles so that he can measure the response of his instrument in terms of numbers of a chemical species. Often he employs methods so sensitive that the precision of his instrument is much greater than the precision achieved by weighing. In these cases, he resorts to the equally fundamental procedure of using a standard solution. A standard solution is defined as one whose composition is accurately known. If macro amounts of a solute, measured with precision, are uniformly dispersed into precisely measured large volumes of solution, a micro-sized sample of solute can be obtained by measuring out a small volume of solution.

EXAMPLE 6.1 A chemist wished to measure the response of an instrument to 1.00×10^{-6} mole of a substance which had a formula weight of 60.0. If he wants a precision of $\pm 1\%$, how can he conveniently obtain his sample?

If he has a balance capable of weighing to ± 0.0001 g, the smallest sized weight he can use will be 0.0100 g to keep within his precision limit. This will represent

$$\frac{0.0100 \text{ (g)}}{60.0 \left(\frac{\text{g}}{\text{mole}}\right)} = 1.67 \times 10^{-4} \text{ mole}$$

If this is dissolved in 1.000 l of solution (easily measured to ± 0.001 l), he then needs only to take

$$\frac{1.00 \times 10^{-6} \text{ (mole)}}{1.67 \times 10^{-4} \left(\frac{\text{mole}}{\text{l}}\right)} = 6.00 \times 10^{-3} \text{ l}$$

The required 6.00 ml can be measured with an ordinary buret to a precision well within the limit of ± 0.06 ml.

chemical equivalence

The simplest methods for measuring an unknown amount of a particular substance in the presence of other materials is to remove it from solution by reaction with a measurable amount of a reagent. For this purpose, a great variety of specific reactions are used. Most of these reactions are one of the types:

Neutralization, e.g., $H_3O^+ + OH^- \rightarrow 2H_2O$

Oxidation-Reduction, e.g., $2Ce^{+4} + C_2O_4^= \rightarrow 2Ce^{+3} + 2CO_2$

Precipitation, e.g., $Ba^{++} + SO_4^= \rightarrow BaSO_4$

Complexation,[1] e.g.,
$Mg^{++} + H_2EDTA + 2H_2O \rightarrow MgEDTA + 2H_3O^+$

The common criterion is that the reagent react "completely" with the one substance sought. We use the quotation marks deliberately, for no chemical reaction goes to completion, it merely reduces the concentration of substance sought to a very small value. (For example, the amount of Ba^{++} ion in equilibrium with the very slightly soluble $BaSO_4$). The completion of the reaction, or more accurately the point at which the amount of reagent added is chemically equivalent to the substance sought, can be detected. Visual indicators or electronic devices commonly are employed.

Another big advantage in using a solution is the obvious one of being able to deliver, and at the same time to measure, just the amount of chemical substance necessary for a reaction. This is the technique of titration. Accurately calibrated volumetric glassware is employed. Pipets commonly are used to deliver precise, predetermined volumes of solution. Burets are employed to measure the amount of a reagent solution required to react with an unknown to the point of chemical equivalence.

[1] "EDTA" stands for *e*thylene *d*iamene*t*etr*a*cetic acid or its disodium salt. is typical of a variety of chelating agents. See C. N. Reilley, *et al, Jour. Chem. 'uc.* **36,** 555 and 619 (1959).

66 the mole concept in chemistry

It is easy to realize, for anyone who has tried to weigh small amounts of a granular substance, that the measuring of volume by means of a buret is not only more convenient, but equally accurate. For this reason, the reagent solutions used by the analytical chemist are made on the basis of volume of solution rather than on the basis of weight of solvent as were those considered in Chapter 5. The concentration of these volumetric reagent solutions usually is expressed on a *molar* scale. The *molarity of a solution* is defined as the number of moles of solute per liter of solution. An alternative term is *formal*. The preference of some for this term is based on their reluctance to use the word "molar" which stands for the number of "gram molecular weights" of solute per liter of solution. Since it is an anachronism to use even the adjective "molecular" when referring to solute such as a salt composed only of ions, the term "formal" implying the number of *gram formula weights* of solute per liter of solution is more attractive. However, since our use of the word "mole" applies broadly, we shall correspondingly use the term *molar* broadly and avoid semantic arguments.

Preparation of a Standard Solution

The usual method of preparing a solution of a desired molarity is to weigh out the required amount of a pure solid substance, and put it into the desired volume of solution.

> **EXAMPLE 6.2** What weight of $CuSO_4 \cdot 5H_2O$ must be taken to make 0.500 l of 0.0100 molar copper(II) ion solution?

We have now introduced one more method of counting number of moles. In this case the number of moles of solute required will be:

$$0.500 \text{ (l)} \times 0.0100 \left(\frac{\text{mole}}{\text{l}}\right) = \text{number of moles of solute} = 0.00500 \text{ mole}$$

Since the weight of one mole of the solid $CuSO_4 \cdot 5H_2O$ is 249.7 g, we will need:

$$\frac{g \, (g)}{249.7 \left(\frac{g}{mole}\right)} = \text{number of moles}$$

Equating these and solving for g, the weight of solid:

$$\frac{g}{249.7} = 0.500 \times 0.0100$$

$$g = 1.248 \, g$$

This weight of solid is then placed in a 500-ml volumetric flask and water added to fill to the mark.

Note that the question specified a molar concentration of the copper(II) ion. In order to obtain the requisite 0.00500 mole of this ion, it was necessary to take 0.00500 mole of $SO_4^=$ ion and 5×0.00500 mole of H_2O of crystallization along with it in the solid. If the anhydrous form of $CuSO_4$ were used to make the solution, the requisite 0.00500 mole of copper(II) ion would be obtained from $0.00500 \text{ mole} \times 159.6 \text{ g mole}^{-1} = 0.798 \text{ g}$ of $CuSO_4$.

It frequently is necessary, as well as convenient, to prepare a primary standard solution at one concentration, and to dilute it further for use. At other times it may be convenient to prepare a solution of an approximate concentration from the commercially available "concentrated" reagent.

EXAMPLE 6.3 A 250-ml sample of 0.20 M hydrochloric acid is to be made by diluting the appropriate amount of the concentrated reagent, 11.7 M. What volume of the latter should be used?

This is typical of all types of "mixture" or "dilution" problems often encountered first as an elementary algebra exercise. We urge the reader to focus on the *chemical* implications; namely, that what is being done is to distribute a required *number of moles* of solute into a different volume of solution. The number of moles of solute in the final solution must be

$$0.250 \text{ (l)} \times 0.20 \left(\frac{\text{mole}}{\text{l}}\right) = \text{number of moles of solute} = 0.050 \text{ mole}$$

This is then equal to the number of moles of HCl required from the unknown volume of concentrated reagent.

$$\underbrace{0.25 \text{ (l)} \times 0.20 \left(\frac{\text{mole}}{\text{l}}\right)}_{\text{desired distribution of a number of solute moles}} = \underbrace{V \text{ (l)} \times 11.7 \left(\frac{\text{mole}}{\text{l}}\right)}_{\text{original distribution of same number of solute moles}}$$

The solution is made up by introducing 4.27 ml of the concentrated solution into a 250-ml volumetric flask and diluting to the mark with water.

Some textbook authors prefer to introduce the term "millimole" (0.001 mole). This has the convenience of allowing the decimal point to be shifted so that volumes expressed in milliliters can be used directly. Accordingly, the relationship stands:

$$\text{number of millimoles of solute} = \text{volume of solution (ml)} \times \text{molarity}$$

Accordingly, this leads to the working equation:

$$\underbrace{\text{volume (ml)}_o \times \text{molarity } (M)_o}_{\text{original undiluted sample}} = \underbrace{\text{volume (ml)}_f \times \text{molarity } (M)_f}_{\text{final diluted sample}}$$

Neutralizations

The determination by volumetric methods of acidic materials is referred to as acidimetry. The complementary term is alkalimetry for measurements made on alkaline or "basic" materials. There is little need for two terms, except to follow the traditionally accepted vocabulary of analytical chemistry. The amount of acid or base is determined by titration with a standard solution of the opposite type to an "end point." This end point is the response of an indicator which has been introduced to "indicate" the point of chemical equivalence. Or it may be the response of an instrument such as a pH meter which can measure the hydrogen-ion (in water solution: hydronium ion) concentration electrically.

We choose to abandon the conventional designation of "gram equivalent weight" and the correlated "normality" scale of concentrations. The gram equivalent weight is, by definition, always some portion of the weight of a mole of the substance:

$$\text{g. eq. wt.} = \frac{\text{weight of one mole}}{\text{number of replaceable hydrogens (or their equivalent)}}$$

and the normality is defined as the number of g. eq. wts. of solute per liter of solution.

The "number of replaceable hydrogens or their equivalent" is a whole number, usually chosen by inspection of the formula. The same is true of the weight of the mole (the gram-formula weight). Hence, we choose to use the mole as the fundamental unit of amount and to work all equivalence problems on this

[2] We are reluctant even to mention the favorite crutch of "liters times molarity," except to ask the reader to recognize that what this always represents is a number of *moles of solute*.

basis. If the acid H_2X is being considered, a mole will contain two g.eq. wts. and correspondingly any solution of the acid will have a normality equal to twice the molarity. It is to be noted that the number relating "normality" and molarity cannot blithely be taken by inspection of the formula. It can be identified only by observing the chemical reaction. For example, the beginner needs either to be told or to do an experiment to know that one mole of acetic acid, CH_3COOH, will react with only one mole of OH^- rather than four.

Neutralizations are but a special case of the topic "Quantitative Relationships in Chemical Reactions" treated in Chapter 4. The chemical equation establishes coefficients which describe the ratio of numbers of moles of reacting species. Superimposed on this fundamental generalization is the special demand of acid-base equivalence as the means of telling what the equation coefficients will be. We have then:

$$\text{moles of } H^+ = \text{moles of } H^+ \text{ acceptor}$$

The most frequently encountered acidic ion will be H^+ (or H_3O^+ in aqueous solution)[3] and the most common basic ion will be the OH^- ion. Any other acidic or basic ions can be translated into these species to represent the fundamental neutralization reaction in water solution. For example, one mole of the basic ions $O^=$ or $CO_3^=$ will be equivalent to two moles of OH^- (or H^+). The coefficient *two* will be revealed by the equation for the reaction being investigated.

The following example[4] serves to make two emphases. The first is that coefficients for an acid-base reaction must be established by experiment, not by inspection of the formula. The

---

[3] Acidity will be represented in the text as H^+ for economy and convenience. However, to keep the reader on his mental toes we shall use H_3O^+ in equations. See C. A. VanderWerf, "Acids, Bases, and the Chemistry of the Covalent Bond," Reinhold Publishing Corp., New York, N. Y., 1961.

[4] The author is indebted to Professor Harold M. State of Allegheny College for describing this experiment done by his elementary chemistry students.

second is that what is thus determined is *chemical equivalence*, not "neutralization" in the sense of making a final solution with the acidity characteristic of pure H_2O ($H^+ = 1 \times 10^{-7} M$).

> EXAMPLE 6.4 A student was provided with 25.0 ml of a $0.107 M$ H_3PO_4. He titrated this with a $0.115 M$ solution of a NaOH to the end point identified by the color change of the indicator, bromcresol green. This required 23.1 ml. He repeated the titration using phenolphthalein indicator. This time his 25.0 ml of $0.107 M$ H_3PO_4 required 46.8 ml of the $0.115 M$ NaOH. What is the coefficient n in the equation:
>
> $H_3PO_4 + nOH^- \rightarrow nH_2O + [H_{(3-n)}PO_4]^{-n}$ for each reaction?

The number of moles of acid ion reacting will be equal to the number of moles of OH^- provided by the measured amount of NaOH solution. However, the equation indicates that the number of moles of H_3PO_4 will be equal to $1/n$ times the number of moles of OH^-. In the first experiment:

$$\underbrace{0.025 \text{ (l)} \times 0.107 \left(\frac{\text{mole}}{1}\right)}_{\text{moles of } H_3PO_4} = \begin{array}{c}\text{moles of} \\ \text{acid-base} \\ \text{reaction}\end{array} = \frac{1}{n} \times \underbrace{0.0231 \text{ (l)} \times 0.115 \left(\frac{\text{mole}}{1}\right)}_{\text{moles of base, } OH^-}$$

$$\frac{1}{n} = \frac{0.025 \times 0.107}{0.0231 \times 0.115} = 1.01; \quad n \approx 1$$

The reaction detected by bromcresol green is:

$$H_3PO_4 + OH^- \rightarrow H_2O + H_2PO_4^-$$

In the second experiment:

$$\frac{1}{n} = \frac{0.025 \times 0.107}{0.468 \times 0.115} = 0.497; \quad n \approx 2$$

The reaction detected by phenolphthalein is:

$$H_3PO_4 + 2OH^- \rightarrow 2H_2O + HPO_4^{-2}$$

> **EXAMPLE 6.5** A student needed to know the concentration of a solution of $Ba(OH)_2$. For his standardization he weighed out 0.2000 g of potassium acid phthalate (weight of one mole = 204.2 g). His titration indicated equivalence at 27.80 ml of $Ba(OH)_2$ solution. What is the molarity of the base? The equation for the reaction is:
>
> $$2KHC_8H_4O_4 + Ba(OH)_2 \rightarrow 2H_2O + 2K^+ + 2C_8H_4O_4^= + Ba^{++}$$
> $$\text{solid} \qquad \substack{\text{in solution as} \\ Ba^{++} + 2OH^-}$$

2 mole of acid, $KHC_8H_4O_4$, react with 1 mole of base, $Ba(OH)_2$. His sample contains:

$$\frac{0.2000 \, (g)}{204.2 \left(\frac{g}{mole}\right)} = 0.000980 \text{ mole of acid,}$$

which his equation predicts will react with

$$\frac{0.000980}{2} = 0.000490 \text{ mole of } Ba(OH)_2.$$

This amount of $Ba(OH)_2$ will be present in

$$0.02780 \, (l) \times M \left(\frac{mole}{l}\right) = \text{mole of base solution.}$$

In summary:

$$\frac{1}{2} \times \frac{0.2000 \, (g)}{204.2 \left(\frac{g}{l}\right)} = 0.02780 \, (l) \times M \left(\frac{mole}{l}\right)$$

$$M = 0.0176 \text{ molarity}$$

An alternative approach to this problem would be to think directly in terms of number of H$^+$ and OH$^-$ as follows:

$$\text{number of moles of H}^+ = \text{number of moles of OH}^-$$

$$\frac{\text{wt. of acid}}{\text{wt. of one mole of acid}} = l \times M \times \begin{array}{c}\text{number of}\\ \text{OH}^- \text{ per mole}\\ \text{of solute}\end{array}$$

$$\frac{0.2000 \text{ (g)}}{204.2 \left(\frac{\text{g}}{\text{mole}}\right)} = 0.02780 \text{ (l)} \times M \left(\frac{\text{mole}}{\text{l}}\right) \times 2$$

The right hand side of this equality provides an example of the completely general approach to all problems involving titration data. The student who is familiar only with the "liters × normality" approach can recognize that his "normality" thus is related to the more fundamental *molarity*.

Chemical equivalence of acid-base then occurs when:

$$l_a \times M_a \times \begin{array}{c}\text{number of re-}\\ \text{acting H}^+ \text{ per}\\ \text{mole of acid}\end{array} = l_b \times M_b \times \begin{array}{c}\text{number of reacting}\\ \text{H}^+ \text{ acceptors per}\\ \text{mole of base}\end{array}$$

The reader will recognize the variety of possible problems to which this relationship is applicable. Any one, but only one of the quantities can be the unknown. Usually, the formulas provide the number of H$^+$ or OH$^-$ per mole of reactant, the volumes are measured, and the reagent identity is known.

EXAMPLE 6.6 The Ba(OH)$_2$ solution standardized as described in Example 6.5 was used to determine the percent by weight of acetic acid (CH$_3$COOH) in a vinegar sample. 10.03 g of vinegar was diluted to 100 ml and a 25.00 ml sample was titrated with the 0.0176 M Ba(OH)$_2$ solution. 34.30 ml was required for equivalence. What is the percent of acetic acid in the vinegar?

The neutralization reaction is:

$$CH_3COOH + OH^- \rightarrow H_2O + CH_3COO^-$$

The number of moles of OH^- provided by the measured amount of $Ba(OH)_2$ solution is:

$$0.03430 \, (l) \times 0.0176 \, (M) \times \underbrace{2}_{\substack{\text{number of } OH^- \text{ per} \\ \text{mole of } Ba(OH)_2}} = 1.208 \times 10^{-3} \text{ mole of } OH^-$$

This is equal to the number of moles of CH_3COOH reacted so that the *weight* of acetic acid in the 25.00-ml titration sample is:

$$\frac{g \, (g)}{60.05 \, \left(\frac{g}{\text{mole}}\right)} = \begin{array}{c}\text{number of}\\ \text{moles}\\ H^+ \text{ or } OH^-\end{array} = 1.208 \times 10^{-3}$$

$$g = 7.254 \times 10^{-2} \, g.$$

The original 10.03 g of vinegar contained 100/25 times as much, or:

$$7.254 \times 10^{-2} \, (g) \times \frac{100}{25.00} = 0.2902 \, g.$$

$$\frac{0.2902}{10.03} \times 100 = 2.90\%$$

Oxidation-Reduction

A second type of reaction which can be used to reduce the concentration of a particular species to a very low value is oxidation or reduction. Total increase in oxidation number of the substance oxidized must equal total decrease in oxidation number of the substance reduced or if the reaction is being interpreted in terms of electrons, the total number of electrons lost by one reactant species must equal the total gained by the other. This is the situation described by the properly balanced redox equation. We shall not here go into the subject of how to balance such equations. Nearly every author of an elementary chemistry textbook advocates a slightly different form of procedure.[5] We pause only to emphasize once again that the complete correctly balanced equation is necessary to serve as a basis for any quantitative interpretation of the reaction. Redox reactions, like acid-base, are merely special cases of the subject treated in Chapter 4. Parallel with the demand of acid-base equivalence, we recognize a principle of redox equivalence serving to establish the equation coefficients.

The following example[6] is analogous to Example 6.4 and serves to make the similar point that equation coefficients cannot be ascertained merely by inspection of the formula.

EXAMPLE 6.7 The MnO_4^- ion in a solution of $KMnO_4$ is an oxidizing agent capable of oxidizing the HSO_3^- ion in a solution of $NaHSO_3$. The oxidation product is $SO_4^=$ (or HSO_4^- in strongly acid solutions). However, the reduction product will vary, depending upon the acidity of the solution. The end point (chemical equivalence) of the titration can be identified by a color change, since MnO_4^- is a brilliant red color. (See reference in Footnote 6 for procedure details.)

A student used 25.00 ml of 0.017 M HSO_3^- as his standard sample. In a strongly acid medium, this required the addition of 16.90 ml of 0.0100 M MnO_4^- solution by titration. In neutral solution, it required 28.60 ml. Assign oxidation numbers to the manganese in each of the products.

---

[5] See R. G. Yalman, *J. Chem. Educ.* **36,** 215 (1959).
[6] See M. J. Sienko and R. A. Plane, "Experimental Chemistry," McGraw-Hill Book Co., Inc., New York, N. Y., 1961, Expt. 33.

The change in oxidation number of S in HSO_3^-, the reducing agent, is the same in both cases:

$$\underset{HSO_3^-}{\overset{+4}{}} \xrightarrow{\text{—— 2 units change ——}} \underset{SO_4^=}{\overset{+6}{}} \text{ (or } \overset{+6}{HSO_4^-}\text{)}$$

Since the number of moles of HSO_3^- ion reacting is

$$0.02500 \text{ (l)} \times 0.0170 \left(\frac{\text{mole}}{\text{l}}\right)$$

Then the number of moles of redox gain is

$$0.02500 \times 0.0170 \times 2 = 0.000850.$$

If the reader finds it necessary to think in terms of electrons, this number will then be the number of moles of electrons *lost* by the reducing agent.

This change must be equal to the number of moles of redox loss (or electrons gained) by the MnO_4^-. We have, then, in highly acid solution:

$$\underbrace{0.02500 \times 0.0170 \times 2}_{\substack{l_{ox} \times M_{ox} \times \text{change} \\ \text{per} \\ \text{mole} \\ HSO_3^-}} = \underbrace{0.01690 \times 0.0100 \times n}_{\substack{l_{red} \times M_{red} \times \text{change} \\ \text{per mole} \\ \text{of} \\ MnO_4^-}}$$

$$\underbrace{}_{\text{moles of redox gain}} = \underbrace{}_{\text{moles of redox loss}}$$

$$n = \frac{0.02500 \times 0.0170 \times 2}{0.01690 \times 0.0100} = 5.03 \approx 5$$

Since the oxidation number of Mn in MnO_4^- is +7, the product must be manganese in the +2 state (actually Mn^{++}). The balanced equation will be: $2MnO_4^- + 5HSO_3^- + 6H_3O^+ \rightarrow 2Mn^{++} + 5HSO_4^- + 9H_2O$. Similarly in the second case:

$$0.02500 \text{ (l)} \times 0.0170 \left(\frac{\text{mole}}{\text{l}}\right) \times 2 = \text{number of moles of redox change} =$$

$$0.02860 \text{ (l)} \times 0.0100 \left(\frac{\text{mole}}{\text{l}}\right) \times q$$

$$q = \frac{0.02500 \times 0.0170 \times 2}{0.02860 \times 0.0100} = 2.97 \approx 3$$

The oxidation number of the manganese in the product must be $+7 - 3 = +4$, (actually MnO_2). The balanced equation will be: $2MnO_4^- + 3HSO_3^- \rightarrow 2MnO_2 + 3SO_4^= + H_3O^+$.

It will be observed that we have used a fundamental expression for redox equivalence which can serve for all analytical calculations based upon redox reactions. We choose to emphasize the molar basis common to both these and acid-base equivalence problems by again omitting "normalities" and "redox equivalent weight" from the discussion. The student who applies his "ml $\times N$" formula to the data of Example 6.7 will have to allow the same solution of MnO_4^- to have two different "normal" concentrations. This paradox is not inherent in considering that moles of oxidizing agent (MnO_4^-) will produce n or q moles of redox change.

EXAMPLE 6.8   25.00 ml of a solution of Fe^{++} was titrated with a solution of the oxidizing agent $Cr_2O_7^=$. 32.45 ml of 0.01530 M $K_2Cr_2O_7$ solution was required. What is the molarity of the Fe^{++} solution? The reaction is:

$$6Fe^{++} + Cr_2O_7^= + 14H_3O^+ \rightarrow 6Fe^{+3} + 2Cr^{+3} + 21H_2O$$

The oxidation of one mole of Fe^{++} involves one mole of redox change, but the reduction of one mole of $Cr_2O_7^=$ involves 6 moles of redox change.

$$0.02500 \text{ (l)} \times M \left(\frac{\text{mole}}{\text{l}}\right) \times 1 = 0.03245 \text{ (l)} \times 0.0153 \left(\frac{\text{mole}}{\text{l}}\right) \times 6$$

$$\underset{\text{oxidation}}{\text{no. of moles of redox change}} = \underset{\text{reduction}}{\text{no. of moles of redox change}}$$

$$M = \frac{0.03245 \times 0.153 \times 6}{0.02500 \times 1} = 0.1192 \text{ molar}$$

EXAMPLE 6.9 A chemist is preparing to analyze samples that will contain no more than 0.500 g of uranium. His procedure calls for preparing the uranium as U^{+4} ion and oxidizing it by MnO_4^- in acid solution.[7]

$$5U^{+4} + 2MnO_4^- + 6H_2O \rightarrow 5UO_2^{+2} + 2Mn^{++} + 4H_3O^+$$

(2 units change on MnO_4^-; 5 units change on U^{+4})

If he wants to react the total U^{+4} sample with a maximum of 50.00 ml of $KMnO_4$ solution, what concentration should he choose?

$$0.500 \text{ g of uranium} = \frac{0.500 \text{ (g)}}{238 \left(\frac{g}{\text{mole}}\right)} = 0.00210 \text{ mole.}$$

The equation indicates that this sample will involve a total number of moles of redox change thus:

$$\underset{\text{oxidation}}{\text{total moles of redox}} = \frac{0.500 \text{ (g)} \times 2}{238 \left(\frac{g}{\text{mole}}\right)}$$

$$\underset{\text{reduction}}{\text{total moles of redox}} = 0.0500 \text{ (l)} \times M \left(\frac{\text{mole}}{\text{l}}\right) \times 5$$

Equating these and solving for the unknown concentration of MnO_4^- gives:

---

[7] C. J. Rodden, Editor: "Analytical Chemistry of the Manhattan Project," McGraw-Hill Book Co., Inc., New York, N. Y., 1950, p. 68.

chemical equivalence

$$\text{molarity of } MnO_4^- = \frac{2 \times 0.5}{238 \times 0.0500 \times 5} = 0.0168 \, M$$

Precipitation

Precipitation reactions are useful as a means of determining the amount of a constituent in solution only if the reagent chosen is specific for one substance sought. For example, if Ba^{++} is in a solution which also contains Ca^{++}, the precipitant cannot be $CO_3^=$ because both $BaCO_3$ and $CaCO_3$ are only slightly soluble. Once the specific reagent is found, a precipitation reaction can be used in two ways. A reagent solution can be used to titrate to an end point as in the previous methods discussed, or the final precipitate can be dried and weighed.

EXAMPLE 6.10 Zinc can be determined volumetrically by the precipitation reaction:

$$3Zn^{++} + 2(4K^+, Fe(CN)_6^{+4}) \rightarrow K_2Zn_3(Fe(CN)_6)_2 + 6K^+$$

A sample of a zinc ore weighing 1.5432 g was prepared for reaction and required 34.68 ml of 0.1043 M $K_4Fe(CN)_6$ for titration. What is the percentage Zn in the ore?[8]

For equivalence we have:

$$\text{number of moles of Zn} \times 2 = \text{number of moles of } K_4Fe(CN)_6 \times 3$$

$$\frac{g \, (g)}{65.38 \left(\frac{g}{mole}\right)} \times 2 = 0.03468 \, (l) \times 0.1043 \left(\frac{mole}{l}\right) \times 3$$

$$g = 0.3547 \, g$$

$$\% \text{ Zn in ore} = \frac{0.3547}{1.5432} \times 100 = 23.00\%$$

---

[8] See R. A. Day, Jr., and A. L. Underwood, "Quantitative Analysis," Prentice Hall, Inc., Engelwood Cliffs, N. J., 1958, p. 175.

80 the mole concept in chemistry

> EXAMPLE 6.11 A chemist has a pure sample of a platinum complex salt which he knows contains Pt, NH_3, and Br. He would like to determine how many Br and NH_3 should be assigned to the complex ion and how many Br are in the salt as ionic Br^-. That is, he wants to know how to write x, y and z in the formula: $(Pt(NH_3)_xBr_y)^{z+}, zBr^-$. He performs the following experiments:
>
> 1. A sample of 0.150 g of the compound ignited and heated to decomposition produced 0.0502 g of platinum metal.
>
> 2. A second 0.150-g sample was dissolved in water and titrated rapidly with 0.0100 M $AgNO_3$ solution. 51.50 ml was required to precipitate all the ionic bromide by the reaction: $Ag^+ + Br^- \rightarrow AgBr$.
>
> 3. A third 0.150-g sample was heated for two hours on a steam bath in a solution to which 0.200 mole of $AgNO_3$ had been added. This precipitated all the bromine (not just the free ionic Br^-) as AgBr. The weight of the precipitate thus produced was 0.1940 g.

On the assumption that the compound was pure, and that only *one* Pt atom is in one mole of the compound, his first experiment told him:

$$\frac{0.150 \text{ g}}{\text{weight of one mole}} = \frac{0.0502 \text{ (g)}_{Pt}}{195.1 \left(\frac{g}{\text{mole}}\right)_{Pt}} = 2.57 \times 10^{-4} \text{ mole}$$

weight of one mole = 584 g

His second experiment told him that his 2.57×10^{-4} mole of compound contained:

$$0.05150 \text{ (l)} \times 0.0100 \left(\frac{\text{mole}}{\text{l}}\right) = 5.15 \times 10^{-4} \text{ mole of } Br^-$$

So the number

$$z = \frac{5.15 \times 10^{-4}}{2.57 \times 10^{-4}} \approx 2$$

The AgBr precipitate in the third experiment amounts to

$$\frac{0.1940 \,(g)}{187.8 \left(\frac{g}{mole}\right)} = 1.031 \times 10^{-3} \text{ mole}$$

So that

$$y + z = \frac{1.031 \times 10^{-3}}{2.57 \times 10^{-4}} \approx 4$$

The number x can then be determined by

$$\text{wt. of one mole} = 584 \,(g) = \frac{(1 \times Pt + x \times NH_3 + 4 \times Br)}{(195.1 + x \times 17 + 4 \times 79.9) \,(g)}$$

$$x = \frac{584 - 195.1 - 319.6}{17} \approx 4$$

The formula is $(Pt(NH_3)_4Br_2)^{++}, 2Br^-$.

Complexation

A great variety of new analytical reagents have come into use in recent years. Many of these are organic materials which have a remarkable selectivity for forming complexes with specific cations. Among the most versatile are EDTA and closely related compounds. They have the advantage of forming very stable complexes so that small amounts of cations can be detected. Furthermore, the conditions of solution, such as pH, temperature, and presence of indicators make it possible to determine more than one constituent in a mixture by means of the same reagent. The following example is a mere suggestion of the vast number of applications of this type of reaction to analytical problems.

the mole concept in chemistry

> EXAMPLE 6.12 10 ml of tap water containing Ca^{++} and Mg^{++} in the presence of HCO_3^- was properly buffered and the indicator murexide added, the sample was diluted and heated to 60°C. Titration with 0.0100 M EDTA solution changed the indicator color at 7.50 ml. This complexed the Ca^{++} only.
>
> A second 10-ml sample was made basic and Erio T indicator added. Titration with 0.01 M EDTA solution changed the indicator color at 13.02 ml. Under these conditions both Ca^{++} and Mg^{++} are complexed.
>
> If the 10 ml of water sample were to be evaporated to dryness, what weight of $CaCO_3 + MgCO_3$ would be formed?[9]

All EDTA complexes are formed on a 1 to 1 basis with dipositive cations. Hence the titration data mean:

$$\text{no. of moles of } Ca^{++} + Mg^{++} = 0.01302 \text{ (l)} \times 0.0100 \left(\frac{\text{mole}}{\text{l}}\right) = 13.00 \times 10^{-5}$$

$$\text{no. of moles of } Ca^{++} = 0.00750 \text{ (l)} \times 0.0100 \left(\frac{\text{mole}}{\text{l}}\right) = 7.50 \times 10^{-5}$$

$$\text{no. of moles of } Mg^{++} = \qquad = 5.50 \times 10^{-5}$$

The combined weight of the carbonates then will be:

$$7.5 \times 10^{-5} \text{ (mole)} \times 100.0 \left(\frac{g}{\text{mole}}\right)_{CaCO_3} +$$

$$5.5 \times 10^{-5} \text{ (mole)} \times 84.3 \left(\frac{g}{\text{mole}}\right)_{MgCO_3} = 12.2 \text{ milligram}$$

[9] Adapted from H. Flaschka, "EDTA Titrations," Pergamon Press, New York, N. Y., 1959, p. 101.

chapter seven

ELECTROCHEMISTRY

Phenomena which demonstrate the interdependence of electrical and chemical concepts are essentially of two types: A flow of electrons through a substance may produce a chemical reaction or a chemical reaction may be manipulated to cause a flow of electrons through some external circuit. The first of these involves the study of electrolysis and conductivity; the second, the measurements of electromotive force. We shall take a brief look at each of them in turn.

Electrolysis

Michael Faraday in 1834 systematized the observations of electrolysis in several laws: The amount of electrode reaction is proportional to the quantity of electricity flowing; the weights of ions discharged are proportional to their atomic weights divided by their charge; and chemically equivalent amounts of reaction occur at both electrodes. What this provides is one more method of counting equal numbers of chemical units. If some standard weight of substance such as silver is deposited by a measured amount of electricity, the resulting quantity of electricity can also be regarded as a standard unit. Once this is measured for a substance which everyone agrees has a single charge, then from atomic weights, ionic charges can be computed. (The reader should recognize that this is the first experimental method for evaluating ion charge. We have perhaps oversimplified the argument by stating Faraday's laws in modern terminology.) Consistency dictates that the standard amount of substance be the mole. The corresponding quantity of electricity is the *faraday*, $\mathfrak{F} = 96{,}490$ coulomb. (This is the electrical charge on one mole of electrons, and the number commonly used to three significant figures is 96,500.) The coulomb is the unit of quantity of electric charge. Ampere is a measure of rate of flow. When one coulomb per second is flowing past a point, the rate is one ampere. The energy can be calculated as the product of the amount (coulombs) × the electrical pressure (volts). One volt-coulomb = one joule.

We can paraphrase Faraday's laws:

$$\frac{q \text{ (coulomb)}}{z\mathfrak{F} \left(\frac{\text{coulomb}}{\text{mole}}\right)} \text{ mole of electrons will produce } n \text{ mole of reaction at an electrode}$$

Where q is the number of coulombs of electricity flowing, n is the number of moles of chemical substance either oxidized or reduced, and z is the charge per ion. This relationship defines electrochemical equivalence in a manner analogous to the relationships in Chapter 6 for acid-base or redox equivalence.

EXAMPLE 7.1 What weight of silver will be deposited at the negative electrode when a current of 0.500 ampere flows for 10.0 minutes through a solution of $Ag^+NO_3^-$?

$$\text{moles of electrons} = \frac{0.500 \left(\frac{\text{coulomb}}{\text{sec}}\right) \times 10.0 \, (\text{min}) \times 60 \left(\frac{\text{sec}}{\text{min}}\right)}{1 \times 96{,}500 \left(\frac{\text{coulomb}}{\text{mole}}\right)} = \frac{g \, (\text{g})}{107.9 \left(\frac{\text{g}}{\text{mole}}\right)}$$

$$g = 0.335 \text{ g of Ag} \qquad \text{moles of Ag}$$

This type of calculation can be applied to any oxidation or reduction reaction which has been performed with the aid of electricity. In every case the equation for the reaction must be known to determine z in the above equation. Then moles of reaction may be calculated if the amount of current is known.

EXAMPLE 7.2 What volume of O_2 will be liberated by the passage of 1.80 amp for 1.50 hr through an aqueous solution of Na^+OH^-? The oxygen is collected over the aqueous solution at 27° C, 735 mm Hg pressure (vapor pressure of H_2O = 27 mm Hg). Consider the oxidation at the electrode to be $4OH^- \rightarrow O_2 + 2H_2O + 4e^-$.

The moles of O_2 gas will be: (see Chapter 2)

$$n = \frac{P \times V}{R \times T} = \frac{\dfrac{735 - 27}{760} \text{ (atm)} \times V \text{ (l)}}{0.0820 \left(\dfrac{1 \text{ atm}}{\text{mole} \,^\circ\text{K}}\right) \times 300 \, (^\circ\text{K})}$$

The electrolysis will produce

$$\frac{q}{z\mathcal{F}} = \frac{1.80 \left(\dfrac{\text{coulomb}}{\text{sec}}\right) \times 1.50 \text{ (hr)} \times 3600 \left(\dfrac{\text{sec}}{\text{hr}}\right)}{4 \times 96{,}500 \left(\dfrac{\text{coulomb}}{\text{mole}}\right)} = 0.0252 \text{ mole of } O_2$$

Equating and solving for V:

$$V = \frac{0.0252 \times 0.0820 \times 300}{0.932} = 0.665 \text{ l of } O_2 \text{ at specified conditions}$$

EXAMPLE 7.3 A constant current flowed for 2.00 hr through a potassium iodide solution oxidizing the iodide ion to iodine ($2I^- \rightarrow I_2 + 2e^-$). At the end of the experiment, the I_2 was titrated with 21.75 ml of 0.0831 M sodium thiosulfate solution ($I_2 + 2S_2O_3^= \rightarrow 2I^- + S_4O_6^=$). What was the average rate of current flow in amperes?

The number of moles of I_2 will be: (see Chapter 6)

$$\underbrace{\frac{0.02175 \text{ (l)} \times 0.0831 \left(\frac{\text{mole}}{\text{l}}\right)}{2}}_{\text{determined by titration}} = \underbrace{\frac{amps \left(\frac{\text{coulomb}}{\text{sec}}\right) \times 2.00 \text{ (hr)} \times 3600 \left(\frac{\text{sec}}{\text{hr}}\right)}{2 \times 96{,}500 \left(\frac{\text{coulomb}}{\text{mole}}\right)}}_{\text{produced by electrolysis}}$$

$$\text{current} = 0.0242 \text{ amp}$$

Conductivity

The flow of electrons through materials can occur by either of two mechanisms. Electronic conduction involves the movement of electrons through an aggregate of stationary atoms or ions as in a metal. Electrolytic conduction involves the movement of material particles bearing an electrical charge (ions) as in aqueous solutions of salts. The passage of current through an electronic conductor does not change its composition. Hence, electronic conduction will not concern us here. However, a study of electrolytic conduction, since the migration of ions is involved, does supply useful chemical information.

When current is passed through an electrolytic conductor, it is possible to measure the resistance in a manner similar to measurements made on electronic conductors. If standard procedures are used, it is then possible to express the reciprocal of this resistance as a *conductance*.

If a chemist measures the electrical resistance of one electrolytic solution and finds it to be less than that of a second, he says that the conductance (proportional to the reciprocal of the resistance) of the first solution is greater. This fact he recognizes to mean that there are more electrons being carried through the first solution per unit time. This applies equally well to the carrying of "electron deficiencies" (positive charges) on ions going in the opposite direction. The value for the conductance measures the passage of total charge (both + and −) past any one spot in the solution. The magnitude of the conductance must depend on three factors: the number of charged particles (ions) in the solution, the magnitude of the charge on each, and the mobility of the ions.

It is important to bear in mind all three of these factors when making comparisons among conductance data for various ionic solutes. The number of ions in the solution can be stated in terms of the molar concentration. However, the second factor, the number of charges per ion must also be considered. For example, the conductance of a 0.001 M solution of NaCl has the value 124 ohm^{-1}, that for a 0.001 M CuSO$_4$ solution is about 220 ohm^{-1}. The major factor responsible for this difference is the fact that whereas the Na$^+$ and Cl$^-$ ions are singly charged, the Cu^{++} and SO$_4^=$ are each capable of carrying twice as much current. This means that the conductance for a 0.001 M NaCl solution must be compared with that of a 0.001/2 = 0.0005 M CuSO$_4$ solution. These solutions are equivalent in terms of moles of electrons being carried. The conductance of a 0.0005 M CuSO$_4$ solution is 115 ohm^{-1}, a figure more nearly comparable to that of the equivalent NaCl solution. It is then possible to develop a sound theoretical basis for interpreting the differences on the basis of the relative mobilities of the corresponding ions. For example, in the case of the ions here considered, the greater coulombic attraction of +2 ions for −2 ions in the solution of Cu^{++}SO$_4^=$ would impede their motion through the solution more than that of +1 ions for −1 ions in the solution of Na$^+$Cl$^-$. Also, the expected differences in hydration would have an effect.

The consequence of this circumstance has been to establish the convention of expressing solute concentrations in electrochemical equivalents per liter of solution rather than in moles per liter of solution for conductance data.[1] Thus the data:

Solute	NaCl	$BaCl_2$	$LaCl_3$
Concentration (molar)	0.00100	0.000500	0.000333
Conductance (ohm^{-1})	124	134	137

are for solutions containing the same number of Cl^- ions and equal numbers of +charges, but not equal numbers of cations.

Conductance data can be used to give an indication of the number of ionic species contained in a given number of moles (calculated on the basis of the formula). The data given above for NaCl, $BaCl_2$, and $LaCl_3$ suggest that if equal *molar* concentrations were investigated, the numbers would be approximately in the ratio of 1:2:3. The values are:

Solute	NaCl	$BaCl_2$	$LaCl_3$
Concentration (molar)	0.001	0.001	0.001
Conductance (ohm^{-1})	124	266	390

The fact that the conductance of a 0.001 M $LaCl_3$ solution is not exactly three times that of a 0.000333 M solution indicates the kind of variation with concentration which occurs. This variation does not invalidate qualitative prediction, however.

This kind of evidence has been used in support of proposed structures for complex compounds. In Example 5.7 we illustrated how colligative property observations could give information on the number of moles of particles set free in solution from one mole of a complex salt. The material $CoCl_3 \cdot 6NH_3$ in 0.001 M solution produced a freezing-point lowering indicating the presence of four particles (presumed to be ions) per mole: $(Co(NH_3)_6)^{+++}$ and $3Cl^-$. When the conductance of this solu-

---

[1] The reader is cautioned to be on guard against inaccurate statements in the textbook literature. In some of the classic inorganic chemistry texts of former years, *molar* conductance figures are incorrectly labelled "equivalent conductivity." In a text bearing a 1960 copyright, though the numbers are correctly called molar conductance, the concentration figures are labelled "normalities." See K. J. Mysels, "Textbook Errors," *Jour. Chem. Educ.* **36**, 303 (1959). However, the discussion is correctly handled in E. J. King, "Qualitative Analysis and Electrolytic Solutions," Harcourt Brace & Co., New York, 1959, p. 109, to cite just one example.

tion (approximately 0.001 M) is measured, it is found to be about 432 ohm^{-1}. Comparison of this figure with those given above suggests that there must be 4 moles of ions per mole of solute. Thus not only the number of particles but the fact that they are charged is established quite conclusively.

Electromotive Force

It was pointed out in Chapter 6 that when an oxidizing agent and a reducing agent are put into the same solution at comparable concentrations, a reaction occurs which goes virtually to "completion." The choice of reagents is based upon this ability of one material to lower its oxidation number (accept electrons) at the expense of the other material's having increased its oxidation number (donated electrons). The varying tendency or ability of a substance to give up electrons, that is, to be oxidized, can be formulated as an "oxidation potential." Although it is impossible to obtain absolute value measurements of this driving force for reaction, it is possible very accurately to measure relative values for a substance in comparison to that of some assigned standard substance. This driving force, an electrical potential, is expressed in volts, the conventional unit of electrical "pressure." Virtually every textbook that includes a discussion of oxidation-reduction or the rudiments of electrochemistry provides a table of standard electrode potentials.

By the use of such a table, it is possible to predict the course of chemical reactions involving the tabulated species. If, for example, a student wants to anticipate which of the following reactions will occur:

$$Sn^{+2} + 2Fe^{+3} \rightarrow Sn^{+4} + 2Fe^{+2} \text{ or}$$

$$2Fe^{+2} + Sn^{+4} \rightarrow 2Fe^{+3} + Sn^{+2},$$

he can consult the table which gives the oxidation potentials:

$$Sn^{+2} \rightarrow Sn^{+4} + 2e^- \qquad E^0 = -0.14 \text{ volts}$$

$$\text{and } Fe^{+2} \rightarrow Fe^{+3} + e^- \qquad E^0 = -0.75 \text{ volts}$$

to tell that the reduced form of tin, Sn^{+2}, will react spontaneously with the oxidized form of iron, Fe^{+3}, as described in the first equation above. If the oxidation reaction, $Sn^{+2} \rightarrow Sn^{+4} + 2e^-$, were to be separated, but put in electrical contact with the reduction reaction, $Fe^{+3} \rightarrow Fe^{+2} + e^-$, with the ion concentrations all at the same effective molarity, an electromotive force of +0.61 volts could be measured in the circuit.

We have chosen the above example deliberately to justify our including this topic at all. The reader will note that the balanced equation involves *two* moles of Fe^{+3} and Fe^{+2}. The oxidation half-reaction quoted from the table of standard electrode potentials is written for only *one* mole of Fe^{+2} and Fe^{+3}. Yet the combination of the $E°$ values $(-0.14 \text{ v}) - (-0.75 \text{ v}) = (0.61 \text{ v})$ is taken as the difference with no regard for the coefficients of the equation or the electrical balance.

The justification for this calculation procedure lies in the fundamental nature of the property being considered. The electromotive force of an electrode reaction is an intensive property, not an extensive one. It is a property characteristic only of the particular species being considered. It does not depend on the total amount of reaction occurring. The voltage developed will be exactly the same if the appropriate measurements are made on drops of solution or on gallons, provided the concentrations (mole l^{-1}) are unchanged. The contrast to an electrolysis reaction should be clear to the reader. The electromotive force in volts expresses a relative *tendency* for reaction; an electrolysis involves an *amount* of reaction. In the former, the potential for reaction is expressed. The mole concept inherently related to *amount* is useful only in connection with the latter where the results of an accomplished reaction are described. If an electrochemical equation is to serve as the basis for an *energy* calculation, both of these factors must be taken into consideration. The electrical energy is expressed as z (moles) $\mathfrak{F}$ (coulombs mole^{-1}) E (volts) = joules.

chapter eight

AVOGADRO'S NUMBER

The mole, thus far in our discussion, has been a definite, constant amount of any chemical species. "Amount" has been measured in various ways, by weighing, by observing P, V, and T for gases, by observing the freezing point of a solution, etc. In every case, the measurement has been interpreted in molar terms by comparison. The behavior of a standard mole was compared with the data at hand to determine how many moles of material were in a sample. Knowing the number of moles, we have been able to describe the behavior of individual atoms, molecules, ions, or electrons as if we had been observing the single particles.

Thus far, it has not been necessary to know how many units are in a mole. The term has meant a constant number mole^{-1} Avogadro's Number, N, yet the actual numerical value has not been needed. The reader can readily see that we have covered many typically chemical topics. (He may add, ". . . without knowing what we were talking about!") It is clear, however, that much additional information can be gained by knowing the value of Avogadro's Number. Many of the experiments now possible with modern apparatus and modern theory to guide the research are truly at the level of discrete atomic particles.[1] Thus it is necessary to know the actual value of Avogadro's Number with high precision.

Many methods are available for making this determination. In principle *any* relationship for a macroscopic measurable quantity which can be interpreted theoretically in terms of the number of particles can serve as a basis for the calculation.[2] The few we shall discuss are chosen for their simplicity and directness. The "directness" in every case means being able to count particles at the atomic level.

Radioactivity

Probably the most convincing direct evidence for the atomicity of matter is the phenomenon of radioactivity. The discrete flashes of light on a fluorescent screen, or clicks of a scaler connected to a Geiger counter each mean a specific atomic event. Rutherford and his contemporaries in 1911 were able to count the number of disintegrations occurring in standard samples of such naturally radioactive materials as radium. They also learned that the alpha particles emitted by radium would pick up electrons and become helium gas molecules. Since one disintegration produced one gas molecule, counting the former and measuring the moles of the latter made it possible for them to accomplish a quite accurate evaluation of N.

EXAMPLE 8.1 Rutherford measured the helium produced in one day by 192 milligrams of radium in equilibrium with its disintegration products Rn, RaA, and RaC. The amount of helium was 0.0824 mm³ (0° C, 1 atm). One gram of the mixture of radium and its daughters emitted 1.36×10^{11} alpha particles per second. Calculate N.[3]

---

[1] The reader may recall, for example, that the discovery of the element mendelevium was accepted by the scientific world after only four atoms had been synthesized. See A. Ghiorso, *et al*, *Phys. Rev.* **90,** 1518 (1955).

[2] The review given by A. A. Sunier, *Jour. Chem. Educ.* **6,** 299 (1929) is a good place for a student to start. He should be on the lookout for arithmetical slips and be aware of changes due to accurate modern numerical values, however.

[3] These figures are quoted from the original report by E. E. Boltwood and E. Rutherford, *Philosophical Magazine (6)*, **22,** 588, 599 (1911).

The number of disintegrations occurring in the 192-milligram sample in one day was:

$$\frac{0.192 \,(g)}{1.000} \times 1.36 \times 10^{11} \left(\frac{disinteg}{sec\,gram}\right) \times 24 \left(\frac{hr}{day}\right) \times 3600 \left(\frac{sec}{hr}\right) =$$

$$2.25 \times 10^{15} \text{ disintegrations}$$

This is also equal to the number of He molecules produced. The number of moles of helium, n, produced in one day is:

$$n = \frac{1.00 \,(atm) \times 8.24 \times 10^{-8} \,(l)}{0.0820 \left(\frac{1\,atm}{mole\,°K}\right) \times 273 \,(°K)} = 3.68 \times 10^{-9} = \text{moles of helium}$$

$$\mathcal{N} = \frac{2.25 \times 10^{15} \,(molec)}{3.68 \times 10^{-9} \,(mole)} = 6.14 \times 10^{23} \text{ molecules mole}^{-1}$$

Electrolysis

In Chapter 7, we repeatedly referred to the faraday, $\mathfrak{F}$, as the amount of electricity carried by the passage of one mole of electrons. The reader can readily recognize that if this definition is valid, and the charge on the individual electron is known, $\mathcal{N}$ can be evaluated. Measuring the charge on the electron was the object of the famous Millikan oil-drop experiment first successfully accomplished about 1909. The earliest values were subject to considerable uncertainty, chiefly because the value for the viscosity of air was not accurately known.[4] (The viscosity would be involved in relating the speed of movement of charged droplets through air in response to the attraction or repulsion of charged plates). However, the charge on the electron is now known with great precision: 1.60209×10^{-19} coulomb. Accordingly, since the chemist is now able to weigh or otherwise measure a mole of chemical reaction and hence the value for the faraday with comparable precision, it is possible to evaluate $\mathcal{N}$ most accurately by the simple relationship:

[4]Compare the results reported by R. A. Millikan in *Phil. Mag.* **34**, 1 (1917), $\mathcal{N} = 6.06 \times 10^{23}$, with those reported earlier in *Phil. Mag.* **19**, 228 (1910), $\mathcal{N} = 6.18 \times 10^{23}$ and *Phys. Rev.* **32**, 396 (1911), $\mathcal{N} = 5.92 \times 10^{23}$.

$$N = \frac{\mathfrak{F}}{e} = \frac{96{,}490 \left(\frac{\text{coulomb}}{\text{mole}}\right)}{1.60209 \times 10^{-19} \left(\frac{\text{coulomb}}{\text{electron}}\right)} = 6.0227(8) \times 10^{23} \text{ electrons mole}^{-1}$$

The reader should note that the number of significant figures here given is sufficient to demand a careful definition of the atomic weight scale (see Chapter 1). The numbers here quoted are based on the latest value of the faraday: $96{,}489.9 \pm 2$ coulomb on the chemical atomic weight scale reported by the National Bureau of Standards.[5] With the new unified scale of atomic weights and the designation of one mole of ^{12}C atoms as 12 g exactly, the value of $\mathfrak{F}$ is $96{,}485.8$ coulombs mole^{-1} Hence, N is:

$$N = \frac{96{,}485.8 \left(\frac{\text{coulomb}}{\text{mole}}\right)}{1.60209 \times 10^{-19} \left(\frac{\text{coulomb}}{\text{electron}}\right)} = 6.02246 \times 10^{23} \text{ electrons mole}^{-1}$$

The following example often is suggested as a method by which a beginning science student can "measure" N. Actually, he measures the ratio of charge on a hydrogen ion to the mass of a hydrogen atom. However, given values for the crucial constants, R and e, N can be calculated from his data.[6]

> **EXAMPLE 8.2** A student collected 50 ml of H_2 gas over water at 23° C, 740 mm Hg barometric pressure. The H_2 was produced by the electrolysis of water. The voltage was constant at 2.1 volts, the current averaged 0.50 amp for 12 min 20 sec. Using the accepted values of R and e, calculate N.

---

[5] D. N. Craig, J. I. Hoffman, C. A. Law, and W. J. Hamer, *Jour. Res. Nat'l. Bur. Standards* **64A**, 381 (1960).
[6] See R. H. Ellis and R. B. Rauch, *Jour. Chem. Educ.* **30**, 460 (1953). The comments in a letter to the editor, *Jour. Chem. Educ.* **31**, 46 (1954) are most pertinent.

The student first uses his data to calculate a value of $\mathfrak{F}$. He has measured a number of moles of chemical reaction (PVT data for H_2 gas). Knowing the electrolysis equation ($2H_3O^+ + 2e^- \rightarrow H_2 + 2H_2O$), he can calculate the number of moles of electrons *chemically*. He can also do this *electrically*, by knowing the amperage and the time (the voltage is not involved). Thus:

chemical counting:

$$n = \text{moles of } H_2 \text{ gas} = \frac{\frac{740-21}{760} \text{ (atm)} \times 0.050 \text{ (l)}}{0.0820 \left(\frac{1 \text{ atm}}{\text{mole }°K}\right) \times 296 \text{ (}°K\text{)}} = 1.95 \times 10^{-3} \text{ mole}$$

electrical counting:

$$n = \text{moles of } H_2 \text{ gas} = \frac{0.50 \left(\frac{\text{coulomb}}{\text{sec}}\right) \times 12.33 \text{ (min)} \times 60 \left(\frac{\text{sec}}{\text{min}}\right)}{2 \times \mathfrak{F} \left(\frac{\text{coulomb}}{\text{mole}}\right)} = \frac{185}{\mathfrak{F}} \text{ mole}$$

$$1.95 \times 10^{-3} = \frac{185}{\mathfrak{F}}$$

$$\mathfrak{F} = 9.50 \times 10^4 \text{ coulombs mole}^{-1}$$

Then $\mathcal{N}$ has the value

$$\frac{\mathfrak{F} \left(\frac{\text{coulomb}}{\text{mole}}\right)}{e \left(\frac{\text{coulomb}}{\text{electron}}\right)} = \frac{9.50 \times 10^4}{1.60 \times 10^{-19}} = 5.94 \times 10^{23} \text{ electrons mole}^{-1}$$

Surface Area of Monomolecular Films

Another method for the estimation of Avogadro's Number, $\mathcal{N}$, is well within the limitations imposed by equipment available to the student of introductory chemistry. This involves the "counting" of molecules which have spread out in a layer assumed to be only one molecule thick on a water surface. The reader is referred to procedures suggested in the literature.[7] The method consists of spreading a known weight of an acid, such as oleic acid, on a water surface so that the area occupied by the acid film can be measured. From the weight of acid and the known weight of a mole of it, the number of moles of acid can be calculated. From the known density of the acid, the volume of the sample can be calculated and hence the thickness of the film. If a reasonable shape for the molecule is assumed (or if known dimensions are available),[8] the number of molecules in the sample can be calculated. Thence, N, the number per mole can be found.

> EXAMPLE 8.3 A student dissolved 0.044 g of oleic acid in 500 ml of pentane. 0.10 ml of this solution occupied 75 cm^2 of surface after the pentane had evaporated. The weight of one mole of the acid was determined to be 286 g. The density of the pure acid was 0.873 g cm^{-3}. Calculate the thickness of the layer. Calculate N on the basis of the assumptions (a) the molecules are cubes, (b) rectangular solids with height = twice the side, (c) cylinders closely packed on end with h = twice the diameter, and (d) using the value 46Å^2 for the surface area of one molecule.

The weight of acid is:

$$\frac{0.044 \times 0.10}{500} = 8.8 \times 10^{-6} \text{ g}$$

---

[7] See L. C. King and E. K. Neilson, *Jour. Chem. Educ.* **35,** 198 (1958) from which the data here given have been quoted; also A. L. Kuehner, *Jour. Chem. Educ.* **19,** 27 (1942).

[8] Many fatty acids exhibit a film area of about 20.5 (Å)^2 per molecule; N. K. Adam, "The Physics and Chemistry of Surfaces," Oxford University Press, London, 1941, p. 47. However, oleic acid occupies 46Å^2 per molecule; see I. Langmuir, *Jour. Am. Chem. Soc.* **39,** 1848 (1917).

The number of moles of acid is:

$$\frac{8.8 \times 10^{-6} \text{ (g)}}{286 \left(\frac{g}{\text{mole}}\right)} = 3.08 \times 10^{-8} \text{ mole}$$

The volume of this sample is:

$$\frac{8.8 \times 10^{-6} \text{ (g)}}{0.873 \left(\frac{g}{\text{cm}^3}\right)} = 1.01 \times 10^{-5} \text{ cm}^3$$

so that the thickness of the layer is:

$$\frac{1.01 \times 10^{-5} \text{ (cm}^3)}{75 \text{ (cm}^2)} = 1.35 \times 10^{-7} \text{ cm}$$

(a) If the molecules are cubes so that the layer is one molecule thick, one mole will contain:

$$\mathcal{N} = \frac{\dfrac{286 \left(\frac{g}{\text{mole}}\right)}{0.873 \left(\frac{g}{\text{cm}^3}\right)}}{(1.35 \times 10^{-7})^3 \left(\frac{\text{cm}^3}{\text{molec}}\right)} = 1.3 \times 10^{23} \text{ molec mole}^{-1}$$

(b) If the molecules are rectangular solids twice as long as they are wide and set on end, one mole will contain:

$$= \frac{328 \text{ (cm}^3)}{\left[\dfrac{1.35 \times 10^{-7} \text{ (cm)}}{2}\right]^2 \times 1.35 \times 10^{-7} \text{ (cm)}} = 5.3 \times 10^{23} \text{ molec mole}^{-1}$$

(c) If the molecules are cylindrical with a length twice the diameter, set on end and closely packed so that only about 90% of the area is covered (the spaces between the cylinder ends amount to about 10% of the total area):

$$\mathcal{N} = \frac{5.3 \times 10^{23}}{0.90} = 5.9 \times 10^{23} \text{ molec mole}^{-1}$$

(d) If the film is assumed to have no vacant areas, the accepted value, 46Å² per molecule, means that the film contains:

$$\text{number of molecules} = \frac{75 \text{ (cm}^2\text{)}}{46 \times 10^{-16} \left(\frac{\text{cm}^2}{\text{molec}}\right)} = 1.6 \times 10^{16} \text{ molec}$$

This is the number of molecules in 3.08×10^{-8} mole. Hence

$$3.1 \times 10^{-8} \text{ (mole)} \times \mathcal{N} \left(\frac{\text{molec}}{\text{mole}}\right) = 1.6 \times 10^{16} \text{ (molec)}$$

$$\mathcal{N} = 5.2 \times 10^{23} \text{ molec mole}^{-1}$$

Crystal Structure Data

A beam of X rays is made up of quanta whose wave length such that the beam can be diffracted by planes of atoms or ions in solids. The well known Bragg relationship, $n\lambda = 2d \sin \theta$, relates the angle of diffraction, θ, with the wave length of the X rays, λ, and the internuclear distance of the units in planes d. n is the order of the diffraction being observed and is an integer. The techniques of using X rays have been so long established that high precision of measurement is now possible. Furthermore, such a large number of substances have been completely characterized that we can speak with assurance about the patterns assumed by atoms or ions in solids almost as if we could "see" them. This offers, then, one more way of counting atomic-sized unit particles.

> **EXAMPLE 8.4** KCl crystallizes in a face-centered cubic unit cell which contains 4K⁺ ions and 4 Cl⁻ ions. The edge of this unit cell measures 6.29082 ± 0.00004 Å (Å = 10^{-8} cm). The density of the crystal is 1.9893 ± 0.0001 g cm^{-3}. What is the value of $\mathcal{N}$?[9]

[9] P. H. Miller, Jr., and J. W. M. Du Mond, *Phys. Rev.* **57**, 206 (1940).

If the unit cell contains $4K^+ + 4Cl^-$, then the volume of four moles of KCl (4×74.553 g, according to the best 1940 atomic weight data when this work was reported) should represent the volume of one mole of unit cells. Hence,

$$\text{volume of 4 mole of KCl} = \frac{4 \times 74.553 \left(\frac{g}{mole}\right)}{1.9893 \left(\frac{g}{cm^3}\right)}$$

$$\text{number of unit cells in this volume} = \frac{4 \times 74.553}{1.9893 \times (6.29082 \times 10^{-8})^3}$$

$$\mathcal{N} = 6.0215 \times 10^{23} \, (K^+Cl^-) \, mole^{-1}$$

The reader can recognize that the precision of this method is limited more by the purity of materials and the absence of flaws in the crystal which might influence its density than it is in "seeing" the spacing between ionic nuclei. Here, too, the precision of measurement is sufficient to give different values depending on the choice of atomic weight scale in defining the mole. A variety of crystals have been measured with high precision, notably $CaCO_3$ in the form of calcite.

chapter nine

CALCULATION OF MOLECULAR QUANTITIES

Once the chemist has a value for Avogadro's Number, he can translate many *molar* quantities into values which stand for observations on the atomic or molecular level. The simplest of these is the weight of a single atom or molecule. This weight always will be the weight of one mole of the substance divided by N. Thus for the O_2 molecule:

$$\text{weight of one } O_2 \text{ molecule} = \frac{32.00 \left(\frac{g}{\text{mole}}\right)}{6.023 \times 10^{23} \left(\frac{\text{molec}}{\text{mole}}\right)} = 5.312 \times 10^{-23} \text{ g}$$

Similarly, the weight of any individual unit can be calculated if the weight of one mole is known. For example, the weight of one H atom will be:

$$\text{weight of one H atom} = \frac{1.008 \left(\frac{g}{\text{mole}}\right)}{6.023 \times 10^{23} \left(\frac{\text{atom}}{\text{mole}}\right)} = 1.673 \times 10^{-24} \text{ g}$$

It is not always possible to evaluate the volume of individual atoms or molecules by following an approach similar to that for finding the weight of a single unit. The mole is defined on a weight basis; hence, division by N gives an unequivocal answer

On the other hand, even though the volume of a mole of particles can be measured very accurately, the shape and arrangement of the particles must be assumed. For solid forms of metallic elements evidence indicates that they can be considered as spherical atoms in contact with each other. Even if there are no irregularities in the packing of the units into a solid, there still are vacant spaces between the spheres. Furthermore, it probably is not correct to speak of the size of an atom or molecule without recognizing that the boundaries must be indefinite. Consequently, what is taken to be the "diameter" of an atom more accurately is the distance between the centers of the nuclei when the atoms are close-packed in the solid. The following example illustrates this kind of calculation:

> **EXAMPLE 9.1** Aluminum metal at 20° C has a density of 2.70 g cm^{-3}. It crystallizes in a face-centered cubic form that represents the close-packing of spheres. In such an arrangement there is 25.9% void space.[1] What is an approximate volume and diameter of an aluminum atom?

One mole of aluminum will have a volume of:

$$\frac{27.0 \left(\frac{g}{\text{mole}}\right)}{2.70 \left(\frac{g}{\text{cm}^3}\right)} = 10.0 \text{ cm}^3 \text{ mole}^{-1}$$

74.1% of this will be occupied by N spheres of diameter d.

$$\frac{4}{3} \pi \left(\frac{d}{2}\right)^3 N = 10.0 \times 0.741$$

---

[1] The reader can verify this as an exercise in solid geometry. The unit cube of a face-centered cubic array, with side a will contain the equivalent of 4 spheres each with radius $a/2\sqrt{2}$. Hence in the solid volume a^3, there will be a volume of $4[\frac{4}{3}\pi(a/2\sqrt{2})^3]$ taken up by the spheres.

$$\left(\frac{d}{2}\right)^3 = \frac{7.41}{1.33 \times 3.14 \times 6.023 \times 10^{23}} = 2.94 \times 10^{-24} \text{ cm}^3$$

$$d = 1.43 \times 10^{-8} \text{ cm}$$

When this type of calculation is applied to data on molecular solids, the values for diameters will be only an approximation. Some molecules such as CCl_4 or CH_4 are nearly spherical, but this would not be expected to hold for diatomic N_2 or O_2.

If we apply this type of calculation to substances in the liquid state, we can expect even less accuracy in the answer. Liquids can be considered as free flowing spheres in contact, but the randomness of their motion must indicate that each unit has an average "sphere of influence" which exceeds its actual dimensions.

Molecules in Gases

In Chapter 2 we demonstrated how the number of moles of a gas can be calculated from observations of the pressure, temperature, and volume: $n = (PV)/(RT)$. We now recognize that n (mole) will contain $\mathcal{N}$ (molec mole^{-1}) so that a simple multiplication is all that is needed to calculate the number of molecules in any gas sample.

EXAMPLE 9.2 Calculate the number of H_2 molecules in the sample of the gas described in Example 2.5.

$$n\mathcal{N} = \frac{6.023 \times 10^{23} \left(\frac{\text{molec}}{\text{mole}}\right) \times \frac{736 - 15}{760} \text{ (atm)} \times 0.0512 \text{ (l)}}{0.0820 \left(\frac{1 \text{ atm}}{\text{mole} \, ^\circ\text{K}}\right) \times 291 \, (^\circ\text{K})}$$

$$n\mathcal{N} = 6.023 \times 10^{23} \times 0.00203 = 1.22 \times 10^{21} \text{ molec}$$

The number 22.414 l (STP) is usually remembered as the volume of one mole of a gas (STP). A comparable number which is used frequently in calculations based on the kinetic molecular theory is the number of molecules in one cubic centimeter of a gas (STP). The validity of Avogadro's hypothesis means that just as 22.414 holds for the ideal gas and is approximately true for *any* gas, the number solved for in the following example is similarly of general applicability.

EXAMPLE 9.3 How many molecules of an ideal gas are in 1.000 cm^3 (STP)?

$$nN = \frac{6.023 \times 10^{23} \left(\frac{\text{molec}}{\text{mole}}\right) \times 1.000 \text{ (atm)} \times 0.00100 \text{ (l)}}{0.0820 \left(\frac{\text{l atm}}{\text{mole °K}}\right) \times 273 \text{ (°K)}}$$

$$nN = 2.69 \times 10^{19} \text{ molec cm}^{-3} \text{ at STP}$$

Recall of this number makes possible the order-of-magnitude estimate of the number of molecules as a function of pressure. Since $nN = [(NV)/(RT)]P$, we see that nN will vary directly with the pressure. For example, a typical "high" vacuum in the laboratory is 10^{-6} mm Hg ($\approx 10^{-9}$ atm). Even under these extreme conditions there will be about 20 billion molecules in each cubic centimeter of the residual gas: $nN = (2 \times 10^{19}) \times (10^{-9}) \approx 2 \times 10^{10}$.

Once the number of molecules in a gas sample is known, it is possible to apply the kinetic molecular theory to calculate other descriptions of molecular behavior. One of these quantities is the velocity of the molecules. When this is known, the number of collisions between molecules in a given volume can be counted, as well as the distance a molecule will travel, on the average, between collisions, (the "mean free path").

We refer the reader to standard textbooks for the derivation of the fundamental equation of the kinetic theory: $PV = 1/3 \, Nmc^2$. This equation confirms the anticipation that the kinetic energy of molecules is proportional to the absolute temperature.

Since $PV = RT$ for one mole and $KE = \text{mass} \times (\text{velocity})^2$:

$$\text{Kinetic Energy} = \tfrac{1}{2}Nmc^2 = \tfrac{3}{2}RT$$

In this equation, m is the mass of a molecule, so the quantity Nm will be the mass of a mole of molecules. c is the "root-mean-square" velocity $(\bar{c}^2)^{\frac{1}{2}}$ of the molecules, and the other symbols have their usual connotation. Because the distribution of molecular velocities indicates a slightly greater number of the faster moving molecules than of the slower ones, the average velocity will differ slightly from the root-mean-square value. However, as suggested by the above equation, c_{ave} will be proportional to $\sqrt{T/M}$ and not depend either on the pressure or the volume of the gas. The relationship is:

$$c_{\text{ave}} = \sqrt{\frac{8RT}{\pi M}}$$

The reader should recognize that this imposes the choice of different units for R than have been used up to this point. Note that dimensional equality results if R is represented as follows:

$$c_{\text{ave}}\left(\frac{\text{cm}}{\text{sec}}\right) = \sqrt{\frac{\left(\dfrac{\text{g cm}^2}{\text{sec}^2 \text{ mole } (^\circ K)}\right) \times (^\circ K)}{\left(\dfrac{\text{g}}{\text{mole}}\right)}}$$

Since g cm^2 sec^{-2} units are ergs, we can use the value for $R = 8.314 \times 10^7$ erg mole^{-1} $(^\circ K)^{-1}$ to calculate the average velocity of the molecules.[2]

EXAMPLE 9.4 What is the average velocity of H_2 molecules at 0° C (273° K)?, at 1000° K? What is the velocity of UF_6 molecules at these temperatures?

[2] See Footnote 2, Chapter 2, p. 15.

calculation of molecular quantities

$$c_{ave} = \sqrt{\frac{8 \times 8.314 \times 10^{-7} \left(\frac{g\,cm^2}{sec^2\,mole\,(°K)}\right) \times 273\,(°K)}{3.14 \times 2.016 \left(\frac{g}{mole}\right)}}$$

average velocity of H_2 molecules $= 1.692 \times 10^5 cm\,sec^{-1}$ (about 3800 miles per hour)

At $1000°\,K$, the average speed of a H_2 molecule will be:

$$c_{ave} = \sqrt{\frac{8 \times 8.314 \times 10^7 \times 1000}{3.14 \times 2.016}} = 3.24 \times 10^5 cm\,sec^{-1}$$

for the UF_6 molecules (weight of one mole = 352 g) at $273°\,K$, the average velocity will be $1.28 \times 10^4 cm\,sec^{-1}$.

Two other descriptions of gaseous behavior are possible once the number of molecules in a mole is known. These are the number of collisions and the distance a molecule travels between collisions. The number of collisions per second between molecules in one cm^3 is predicted by the theory to be:

$$Z = \tfrac{1}{2} \sqrt{2}\,\pi\,(nN)^2 d^2 c_{ave}$$

and the mean free path:

$$\lambda = \frac{1}{\sqrt{2}\,\pi\,(nN) d^2}$$

Here (nN) is the number of molecules in one unit vol., and d is the diameter of the molecule. We will not go into the details either of derivation or calculation. However, it may be interesting for the reader to recognize the interrelationship between the various molecular quantities thus far discussed. At STP conditions, we found that (nN) is of the order of 10^{19} (molec per cm^3) and varies directly with the pressure. Molecular diameters are of the order of 10^{-8} cm. Average velocities are of the order of $10^5 cm\,sec^{-1}$. We have then at STP if we combine all the numerical factors (π, etc.) into the number 10:

number of collisions
in 1 cm³ in 1 sec $\approx 10 \times [10^{19} \text{ (cm}^{-3})]^2 \times [10^{-8} \text{ (cm)}]^2 \times \left[10^5 \left(\frac{\text{cm}}{\text{sec}}\right)\right]$
STP

$$Z_{STP} \approx 10^{28}$$

Even at the very low pressure of a good laboratory vacuum ($\approx 10^{-9}$ atm), the number of collisions will be about ten billion second^{-1} cm^{-3}.

$$Z_{10^{-9} \text{ atm}} \approx 10 \times (10^{10})^2 \times (10^{-8})^2 \times (10^5) \approx 10^{10}$$

In a similar manner, an estimate of λ, the mean free path of molecules at STP, will be:

$$\lambda_{STP} \approx \frac{1}{10 \times [10^{19} \text{ (cm}^{-3})] \times [10^{-8} \text{ (cm)}]^2}$$

$$\lambda_{STP} \approx 10^{-4} \text{ cm}$$

Since the mean free path varies inversely as the number of molecules, we find that in the good laboratory vacuum ($\approx 10^{-9}$ atm) the molecules can travel 10^9 times as far as they can at STP or about 10^5 cm between collisions.

Molecular Energies

It is beyond the scope of this discussion to examine in detail how the chemist is able to interrelate his information about the energy exchanges which accompany all chemical reactions. The reader's experience and his common-sense extrapolation of the theme of this book will lead him to recognize the importance of the *mole* as the basis for all energy comparisons. Thus we find that the tables of thermochemical data define quantities as the "heat of formation per mole" or the "heat of combustion per mole." A great mass of accurate calorimetric data is now available so that the heat of reaction can be calculated for chemical processes which would be difficult or impossible to observe directly; for example, the heat of formation of CH_4 from diamond and H_2 gas. For the reaction: C (diamond) + $2H_{2(g)} \rightarrow CH_{4(g)}$, the molar heat of formation of $CH_{4(g)}$ is -17.865 Kcal (the minus implies the evolution of heat). The heat evolved when 3 mole of CH_4 are formed is 3×-17.865 Kcal; for 1.00 g of CH_4 it is $1/16 \times -17.865$ Kcal etc.

It is appropriate, however, as a final section in the elaboration of the mole concept theme briefly to indicate how the modern chemist uses it to gain detailed information about molecules from observations of energy exchanges.

It is easily recognized that atoms or molecules can "hold" energy in various ways. We have already described one as the average kinetic energy of gas molecules. This is the energy of motion they possess as a consequence of their being at a given temperature. As long as a gas sample is in thermal contact with its environment at constant temperature, the kinetic energy of its molecules will be proportional only to T.

$$\tfrac{1}{2} \mathcal{N} m c^2 = \tfrac{3}{2} RT$$

If we choose to speak of the kinetic energy of one molecule, we can do so only with reservations. Observations made on large numbers of molecules reflect some kind of an average quantity which must be treated statistically to be interpreted as a value for the individual molecule. These reservations are handled by the theory which takes into account the distribution of velocities. Hence, it is common practice to speak of the "kinetic energy of a molecule" as:

$$\mathrm{KE} = \frac{3}{2} \times \frac{R}{\mathcal{N}} \times T$$

We note that since R has dimensions of energy mole^{-1} (°K)$^{-1}$, division of R by $\mathcal{N}$ will retain the same energy dimensions but be a value per molecule. This quantity $R/\mathcal{N}$ is an important and useful constant known as the Boltzmann constant and designated by k.

$$k = \frac{R}{\mathcal{N}} = \frac{8.314 \times 10^7 \left(\frac{\text{erg}}{\text{mole}\,°\text{K}}\right)}{6.023 \times 10^{23} \left(\frac{\text{molec}}{\text{mole}}\right)} = 1.380 \times 10^{-16} \text{ erg } (°\text{K})^{-1}\text{molecule}^{-1}$$

EXAMPLE 9.5 Express the kinetic energy of a gas at 300° K in terms of cal mole^{-1} and erg molec^{-1}. The mechanical equivalent of heat is 4.184×10^7 erg = 1 cal.

At 300° K, we have

$$\text{KE} = \frac{3}{2} RT = \frac{3}{2} \times \frac{8.314 \times 10^7 \left(\frac{\text{erg}}{\text{mole}\,°\text{K}}\right) \times 300\,(°\text{K})}{4.184 \times 10^7 \left(\frac{\text{erg}}{\text{cal}}\right)}$$

$$\text{KE} = 1.50 \times 1.99 \left(\frac{\text{cal}}{\text{mole}\,°\text{K}}\right) \times 300\,(°\text{K}) = 894 \text{ cal mole}^{-1}$$

which is the same as:

$$\text{KE} = 1.50\,kT = 1.50 \times 1.380 \times 10^{-16} \left(\frac{\text{erg}}{\text{molec}\,°\text{K}}\right) \times 300\,(°\text{K})$$

$$\text{KE} = 6.21 \times 10^{-14} \text{ erg molec}^{-1}$$

The typical calculations involving the energies of particles accelerated by electrical forces usually express the energy in electron volts. An electron volt is the energy imparted to an electron, (charge = 1.602×10^{-19} coulomb) when it falls through a potential of 1.00 volt. It is useful to be able to translate ev per particle into calories or Kcal per mole.

$$\frac{1.00 \text{ ev}}{\text{per particle}} = \frac{1.602 \times 10^{-19}\,(\text{coul}) \times 1.00\,(\text{v}) \times 6.023 \times 10^{23} \left(\frac{\text{particles}}{\text{mole}}\right)}{4.184 \left(\frac{\text{volt coulomb}}{\text{cal}}\right) \times 10^3 \left(\frac{\text{cal}}{\text{Kcal}}\right)}$$

1.00 ev per particle = 23.05 Kcal mole^{-1}

k has the value 8.62×10^{-5} ev $(°\text{K})^{-1}$ molecule^{-1}.

calculation of molecular quantities

EXAMPLE 9.6 When the ^{235}U nucleus undergoes fission, the energy released imparts about 160 Mev (10^6ev) kinetic energy to the fission fragments. Also several neutrons are released with about 2 Mev of energy each. What is the equivalent of fission energy in Kcal mole^{-1}? What is the effective "temperature" of the neutrons? Estimate the speed of these neutrons.

The above conversion factor can be applied directly to calculate the fission equivalent energy as:

$$\text{one mole of fission} = 1.60 \times 10^8 \left(\frac{\text{ev}}{\text{particle}}\right) \times 23.05 \left(\frac{\text{Kcal}}{\text{mole}} \times \frac{\text{particle}}{\text{ev}}\right) =$$
$$3.69 \times 10^9 \text{ Kcal mole}^{-1}$$

The kinetic energy of the neutrons will be:

$$\text{KE} = 2.0 \times 10^6 \left(\frac{\text{ev}}{\text{particle}}\right) \times 23050 \left(\frac{\text{cal}}{\text{mole}} \times \frac{\text{particle}}{\text{ev}}\right)$$

$$\text{KE} = 4.61 \times 10^{10} \text{ cal mole}^{-1}$$

This is equivalent to a mole of gas molecules at a temperature represented as:

$$\text{KE} = \tfrac{3}{2}RT$$

$$4.61 \times 10^{10} \left(\frac{\text{cal}}{\text{mole}}\right) = \frac{3}{2} R \left(\frac{\text{cal}}{\text{mole }°\text{K}}\right) \times T_\text{eff}\,(°\text{K})$$

$$T_\text{eff} = \frac{4.61 \times 10^{10}}{1.99 \times 1.5} = 1.54 \times 10^{10}\,°\text{K}$$

The speed of the neutrons can be estimated with the relationship used in Example 9.5. The mass of one mole of neutrons is 1.009 g. Hence we have:

$$(c_{ave})^2 = \frac{8 \times 8.314 \times 10^7 \left(\frac{\text{g cm}^2}{\text{sec}^2\text{mole}(°\text{K})}\right) \times 1.54 \times 10^{10} \, (°\text{K})}{3.14 \times 1.009 \left(\frac{\text{g}}{\text{mole}}\right)}$$

$$c_{ave} = 1.8 \times 10^9 \text{cm sec}^{-1}$$

Molecules can "hold" energy by other means than the kinetic energy of their motion. If two or more atoms are combined into a molecule, the fact that the centers of the nuclei do not coincide with the "geographical" center of the molecule means that it will have a moment of inertia and hence can absorb energy to set it rotating. Also, whenever more than one atom with its massive nucleus is present, these nuclei can vibrate with respect to each other and thus absorb energy. Sometimes so much vibrational energy is absorbed by the molecule that the atoms can break away from each other; the molecule dissociates into fragments. Still an additional mode of energy absorption is possible; electrons in the molecule can be excited to higher energy levels or leave the molecule entirely to produce an ion.

The absorption of energy from an external source by molecules and the complementary emission of energy as an excited molecule returns to its lower level of potential energy is quite adequately understood and interpreted by the use of the quantum theory. The fundamental relationship is that equating the energy of a quantum ("unit bundle" of electromagnetic radiation) with its frequency or wave length:

$$E = h\nu = hc/\lambda$$

Here ν is the frequency (sec^{-1}), h is Planck's constant, 6.624×10^{-27} (erg sec), c is the velocity of light, 2.9979×10^{10} (cm sec^{-1}), and λ is the wave length (cm) often expressed in Angstrom units (10^{-8}cm). This relationship enables the spectroscopist to measure wave lengths of absorbed or emitted light and translate the information into a pattern of energy levels for the molecule. Thus we see that with spectroscopy supplying the raw data and the quantum theory the method for inter-

calculation of molecular quantities 111

preting it, a great deal of information about individual molecules is now available.

It is consistent with our theme to see how it is possible for the chemist to correlate the energetic behavior of individual molecules as revealed by spectroscopy with energy relationships on the macroscopic or molar level.

> **EXAMPLE 9.7** A common source of ultraviolet light is the quartz mercury lamp which produces light of 2537 Å. What is the energy of one quantum? What does this energy correspond to in Kcal mole^{-1}, i.e., for Avogadro's Number of quanta?

$$E = \frac{h \text{ (erg sec)} \, c \left(\frac{cm}{sec}\right)}{\lambda \text{ (cm)}} = \frac{6.624 \times 10^{-27} \times 3.00 \times 10^{10}}{2.537 \times 10^{-5}}$$

$$E = 7.84 \times 10^{-12} \text{erg per quantum}$$

For one mole of quanta we have:

$$E = \frac{7.84 \times 10^{-12} \left(\frac{erg}{quantum}\right) \times 6.023 \times 10^{23} \left(\frac{quantum}{mole}\right)}{4.184 \times 10^7 \left(\frac{erg}{cal}\right) \times 10^3 \left(\frac{cal}{Kcal}\right)}$$

$$E = 113 \text{ Kcal mole}^{-1}$$

The following table provides a consolidated picture of the relative magnitudes of energies in various portions of the electromagnetic spectrum. An exercise for the reader is to fill in the missing numbers.

Radiation	Wave length (Å)	Frequency (sec^{-1})	E per quantum (erg)	E per mole (Kcal)
X ray	1.	3×10^{18}	2×10^{-8}	3×10^5
U.V.	2000	---	1×10^{-11}	1.5×10^2
Visible	5000	6×10^{14}	---	60
I.R.	---	3×10^{13}	2×10^{-13}	3
Radar	---	3×10^{10}	2×10^{-16}	---

Photochemistry is the study of chemical reactions induced by the absorption of radiant energy. One of the basic assumptions of photochemistry was that proposed first by Einstein: quanta are integrally absorbed by molecules. It follows, then, that the energy of Avogadro's Number of quanta will be an important figure. This is referred to as an *einstein* of energy. The value of an einstein of energy is:

$$E = \frac{Nhc}{\lambda} = \frac{2.858 \times 10^5}{\lambda_{(\text{Å})}} \text{ Kcal mole}^{-1}$$

(The reader may verify this by inserting the proper numerical values and conversion factors.)

The numbers in the last column of the above table are the approximate values for the einsteins of radiation in the corresponding spectral ranges. When comparisons are made between these values and the heats of chemical reactions (usually between 10 and 1000 Kcal mole^{-1}), it is apparent why most photochemical investigations are made with ultraviolet light. It is also apparent that since any chemical reaction between molecules must involve the breaking and reforming of chemical bonds, bond strengths must be about 10 to 100 Kcal mole^{-1} in energy. Reference to the table then suggests that bond breaking probably occurs when molecules absorb quanta of visible or ultraviolet radiation. The shorter the wave length of the quantum, the more energy it carries and the stronger the bond it can disrupt.

It is also apparent from this discussion that the absorption of long-wave-length visible light and infra-red light seldom will break bonds. Rather the molecules absorb these quanta by increasing molecular vibrations and rotations. It is interesting to note that these forms of energy absorption are still slightly more energetic than the kinetic energies of free-moving molecules at normal temperatures. At room temperature, the kinetic energy of a mole of gas molecules will be less than 1 Kcal mole^{-1}, see Example 9.5.

SUMMARY

The number of moles of a substance can be calculated by various means. Choice of method depends on the data provided.

For solids or liquids: $$n = \frac{\text{weight}}{\text{weight of one mole}}$$

For gases: $$n = \frac{\text{pressure} \times \text{volume}}{R \times \text{temperature}}$$

For solutes in volumes of solution: $$n = \text{volume (liters)} \times \text{molarity}$$

For solutes in G grams of solvent: $$n = \frac{G \times \Delta T_f}{K_f \times 1000}$$

For products of electrolysis: $$n = \frac{\text{current (amps)} \times \text{time (sec)}}{z \text{ (charge per ion)} \times \mathfrak{F}}$$

INDEX

Arrhenius, 60
Association, 61
Atom diameter, 101
Atomic weight scales, 5
   comparisons, 9
Atomic weights, 4
Avogadro's hypothesis, 11, 23, 103
Avogadro's Number, 4, 9, 28
   methods of measuring, 91 ff
   numerical value, 93, 94, 99

Berthollet, 27
Berthollide compounds, 27
Berzelius, 5
Boltzmann constant, 107, 109

Cannizzaro, 23
Carbon-12 atomic weight scale, 8
Colligative properties, 52 ff
Complex compounds, 59, 80, 88
Complexation, 81 ff
Conductance, 86
   molar, 88
Cryoscopy, 55
Crystal structure, 98
   cubic close packing, 101

Dalton, 5, 22, 27
Dilution problems, 68
Dissociation, 60
Du Long and Petit, 6
Dumas procedure, 19

"EDTA," 65, 81
Einstein, 112
Electrochemical equivalence, 84, 95
Electrolysis, 83 ff, 93
Electromotive force, 89
Electron volt, 108
Energy, molecular, 106 ff
   kinetic, 104
   quantum, 112
Equation of state, 16
Extensive properties, 48

Face-centered cubic crystal, 101
Faraday, value of, 84, 94
Faraday's Laws, 83
Formal solutions, 66
Formula weight, 5, 66
Formulas, 24 ff, 28 ff

Gay Lussac's Law, 11, 45
Gram equivalent weight, 69
   redox, 77
Gram formula weights, 66
Gravimetric factors, 35

Ideal gas, 16, 103
Intensive properties, 49

Kinetic energy, 104, 107
Kinetic molecular theory, 11, 103

Mean free path, 103, 105
Mechanical equivalent of heat, 108

Millimole, 43, 68
Molality, 53 ff
Molar gas volume, 14, 16, 102
Molar solutions, 66
Mole, definition, 4, 9, 10, 12
Mole-fraction, 20, 46, 51, 54
Molecular weight, 5, 13
Molecules in gases, 102
Monomolecular films, 95

Neutralization, 69 ff
Non-ideal solutes, 58 ff
Non-stoichiometric compounds, 27
Normality, 69
Nuclear fission, 109
Number of collisions, 103, 105
Number of molecules, 102

Oxidation potential, 89
Oxidation-reduction, 74 ff, 85

Photochemistry, 112
Pound-mole, 5, 34
Precipitation, 79 ff

Quantum, 110

"R", gas constant, 15 ff, 18, 94, 104
Radioactivity, 92 ff
Rankine temperature, 18
Raoult's Law, 50 ff
Rast method, 56
Rutherford, 92

Solutions, 47
Standard sample, 63
Standard solution, 64, 66

Titration, 65

Vapor pressure, 50
Velocity of gas molecules, 103
   average, 104
   root mean square, 104

Yields of reaction, 37

SELECTED VALUES OF 1961 ATOMIC WEIGHTS

SELECTED VALUES OF 1961 ATOMIC WEIGHTS

Element	Symbol	Weight	Element	Symbol	Weight
Aluminum	Al	26.9815	Magnesium	Mg	24.312
Antimony	Sb	121.75	Manganese	Mn	54.9380
Argon	Ar	39.948	Mercury	Hg	200.59
Arsenic	As	74.9216	Molybdenum	Mo	95.94
Barium	Ba	137.34	Neon	Ne	20.183
Beryllium	Be	9.0122	Nickel	Ni	58.71
Bismuth	Bi	208.980	Nitrogen	N	14.0067
Boron	B	10.811	Oxygen	O	15.9994
Bromine	Br	79.909	Palladium	Pd	106.4
Cadmium	Cd	112.40	Phosphorus	P	30.9738
Calcium	Ca	40.08	Platinum	Pt	195.09
Carbon	C	12.01115	Potassium	K	39.102
Cerium	Ce	140.12	Rhenium	Re	186.2
Cesium	Cs	132.905	Rubidium	Rb	85.47
Chlorine	Cl	35.453	Scandium	Sc	44.956
Chromium	Cr	51.996	Selenium	Se	78.96
Cobalt	Co	58.9332	Silicon	Si	28.086
Copper	Cu	63.54	Silver	Ag	107.870
Fluorine	F	18.9984	Sodium	Na	22.9898
Gallium	Ga	69.72	Strontium	Sr	87.62
Germanium	Ge	72.59	Sulfur	S	32.064
Gold	Au	196.967	Tantalum	Ta	180.948
Hafnium	Hf	178.49	Tellurium	Te	127.60
Helium	He	4.0026	Thallium	Tl	204.37
Hydrogen	H	1.00797	Thorium	Th	232.038
Indium	In	114.82	Tin	Sn	118.69
Iodine	I	126.9044	Titanium	Ti	47.90
Iron	Fe	55.847	Tungsten	W	183.85
Krypton	Kr	83.80	Uranium	U	238.03
Lanthanum	La	138.91	Vanadium	V	50.942
Lead	Pb	207.19	Zinc	Zn	65.37
Lithium	Li	6.939	Zirconium	Zr	91.22